D0808671

RESULTS FROM THE

SIXTH MATHEMATICS ASSESSMENT

OF THE

NATIONAL ASSESSMENT OF EDUCATIONAL PROGRESS

edited by

Patricia Ann Kenney
University of Pittsburgh
Pittsburgh, Pennsylvania

Edward A. Silver
University of Pittsburgh
Pittsburgh, Pennsylvania

National Council of Teachers of Mathematics

WILLIAM WOODS UNIVERSITY LIBRARY

Copyright 1997 by
THE NATIONAL COUNCIL OF TEACHERS OF MATHEMATICS, INC.
1906 Association Drive, Reston, VA 20191-1593
All rights reserved

Library of Congress Cataloging-in-Publication Data:

Results from the Sixth Mathematics Assessment of the National
    Assessment of Educational Progress / edited by Patricia Ann Kenney,
    Edward A. Silver.
            p.   cm.
        Includes bibliographical references (p.   –   )
        ISBN 0–87353–429–8
            1. Mathematics—Study and teaching—United States—Evaluation.
    2. Mathematical ability—testing.   3. National Assessment of
    Educational Progress (Project)   I. Kenney, Patricia Ann.
    II. Silver, Edward A., 1941–   .
    QA13.R484   1997
    510'.71'273—dc21

                                                            97—5290
                                                                 CIP

This material is based on work supported by the National Science Foundation
under Grant No. RED–9453189. The Government has certain rights in this
material. Any opinions, findings, and conclusions or recommendations
expressed in this material are those of the author(s) and do not necessarily
reflect the views of the National Science Foundation.

The publications of the National Council of Teachers of Mathematics present a variety
of viewpoints. The views expressed or implied in this publication, unless otherwise
noted, should not be interpreted as official positions of the Council.

Printed in the United States of America

QA
13
.R484
1997

Contents

FOREWORD

Mathematics has taken an increasingly important role in everyday life. In this age of information and technology, society's expanding use of data makes it imperative for all citizens to have an understanding of, and facility with reasoning about, quantities. In order to ensure that every child is prepared for full participation in society, teachers, administrators, school board members, and other policymakers must be well informed regarding what children know and can do in school mathematics so that they can use this information to improve mathematics education.

For more than 20 years the National Assessment of Educational Progress (NAEP) has provided data to fill this need. For each NAEP assessment, the National Center for Education Statistics (NCES) publishes a variety of reports based on student performance results and responses to the various questionnaires. Although the information in the NCES-sponsored reports is valuable and useful, those documents merely *report* results and do not interpret the data in terms of the current state of mathematics education in this country. Moreover, the NAEP reports do not speak directly to the needs of classroom teachers and those who engage in the preparation of teachers and the professional development of teachers. Producing interpretive reports of NAEP results has been a long-standing interest of the National Council of Teachers of Mathematics (NCTM) since the first NAEP mathematics assessment, and this volume continues that tradition. With the generous support of the National Science Foundation (NSF) and with additional support from NCES, the writers of this volume have produced what I believe will be a useful interpretation of the results of the sixth assessment of mathematics.

NCTM is an organization of more than 125,000 members, including nearly 40,000 institutional members. It represents teachers, college mathematics educators, supervisors, and other administrators across the United States and Canada. The organization has been primarily responsible for the major movement in curriculum change and instructional renewal in mathematics for the past ten years through the three sets of Standards dealing with curriculum and evaluation, professional teaching, and assessment. The groups responsible for the development of the framework and items for the sixth NAEP mathematics assessment drew upon a draft version of the NCTM *Curriculum and Evaluation Standards for School Mathematics* during their deliberations. Thus, there is an important link between the vision for school mathematics and the content and format of this NAEP mathematics assessment and those that will follow. Because of the link between the NCTM *Standards* and the NAEP assessment, the mathematics education community can make use of the information in this volume to help them in redesigning their curricula, instruction, and assessment as they make progress toward meeting the *Standards*.

I thank the members of the NCTM/NAEP interpretive team for sharing with us their views of the results of the sixth assessment. Their chapters make it clear that some progress has been made toward meeting the goals of the NCTM *Standards* but that much remains to be done. The valuable insights in this volume should help teachers, supervisors, college professors, and policymakers find ways to ensure that all children will be mathematically powerful in the twenty-first century.

JACK PRICE
Past President, NCTM

PREFACE

The National Council of Teachers of Mathematics (NCTM) has a history of preparing and disseminating interpretive reports based on results from mathematics assessments conducted by the National Assessment of Educational Progress (NAEP). For the sixth assessment, NCTM continued its dedication to producing interpretive reports of NAEP results by appointing a steering committee to plan and to secure funds from the National Science Foundation (NSF) for producing the interpretive reports. Members of this steering committee were John A. Dossey, Douglas A. Grouws, Patricia Ann Kenney, Vicky L. Kouba, Mary Montgomery Lindquist, Jack Price, and Edward A. Silver. As a result of the steering committee's efforts, a grant was received from NSF, with supplemental funds provided by the National Center for Education Statistics (NCES), to support the work of a team of writers of interpretive reports for the sixth NAEP mathematics assessment. The project is directed by Edward A. Silver and Patricia Ann Kenney, and the other members of the team are Glendon W. Blume, David S. Heckman, Vicky L. Kouba, Marilyn E. Strutchens, and Judith S. Zawojewski. These members of the interpretive team are the principal authors of the chapters in this monograph. John A. Dossey, Ina V. S. Mullis, and Mary Montgomery Lindquist have also contributed chapters.

The interpretive reports in this volume are based on NAEP findings regarding: (a) the cognitive performance of students at grades 4, 8, and 12 on multiple-choice, regular constructed-response, and extended constructed-response items; (b) students' responses to a variety of background questions dealing with their feelings and beliefs concerning mathematics and their participation in various forms of classroom activity; and (c) teachers' responses to various background questions dealing with the nature of their mathematics instruction. The results are summarized for the different grade levels and subgroups of students by gender and race/ethnicity.

The data appearing in this volume are those that were released by NCES in 1993. Since that release, the NAEP data were revised because of minor technical errors that resulted in slight adjustments in average scale scores and the percents of students who attain the achievement levels as set by the National Assessment Governing Board. For 1992, the average scores were raised about one point, and the revised achievement-level results generally show slightly smaller percents of students at each level than were originally reported. Because the differences between the revised data and the data used in this volume were small and because the revised data were not available to the public until the book was in the final stages of publication, the scale scores and achievement-level data appearing in the chapters that follow are those released in 1993. In our judgment, these small changes in no way affect the interpretations made herein.

We wish to thank all who helped in the preparation of the volume: NAEP representatives at NCES and Educational Testing Service (the NAEP contractor) for their cooperation with the project, the staff at National Computing Systems for facilitating access to the test booklets so that samples of student responses could be gathered, the editorial staff at NCTM for extensive editorial assistance and for publishing the reports, and the people at NCES and NSF for funding the project. We are also grateful to Cengiz Alcaci, a doctoral student at the University of Pittsburgh who served as the project's research assistant, and to the members of the project's advisory board—Jane Armstrong, Joan Countryman, John Dossey, Joan Leitzel, Mary Lindquist, Robert Linn, and Jack Price—for their guidance and support. We thank Jane, John, Mary, and Jack for also serving as reviewers of many of the chapters.

1

Learning about NAEP:
Information concerning the Sixth
Mathematics Assessment

Patricia Ann Kenney

THE NATIONAL ASSESSMENT of Educational Progress (NAEP) is a congressionally mandated survey of the educational achievement of American students and changes in that achievement over time. Since 1969, NAEP
has been assessing what American students know and can do in a variety
of curriculum areas, including mathematics. To provide context for the
achievement results, NAEP additionally collects demographic, curricular,
and instructional background information from students, teachers, and
school administrators. From its inception, NAEP has used national samples
of 9-, 13-, and 17-year-old students selected according to a stratified ran-
dom sampling design. In 1983, NAEP expanded the national samples to
include grade-level results, with the most recent assessment targeting students in grades 4, 8, and 12. In 1988, Congress added a new dimension to
NAEP by authorizing on a trial basis voluntary participation in state-level
assessments.

To date, there have been seven NAEP mathematics assessments administered at the national level. These took place in the school years ending in
1973, 1978, 1982, 1986, 1990, 1992, and 1996, with state-level mathematics
assessments given at grade 8 in 1990 and grades 4 and 8 in 1992 and 1996.
This report examines the results for the sixth mathematics assessment conducted in 1992.[1] To avoid possible confusion for readers, it was decided

---

[1]Previous interpretive reports of NAEP results published by NCTM have referred to the
assessments according to their ordinal designations (e.g., the *fourth* mathematics
assessment). However, NAEP reports published by the National Center for Education
Statistics more commonly refer to the assessment by the year of administration (for
example, the *1986* mathematics assessment). Because both forms have been used
interchangeably in this volume, readers should keep in mind that the fifth assessment is
synonymous with the 1990 assessment, and the sixth assessment, with the 1992
assessment.

that this volume should focus on the national-level findings from the sixth assessment, with far less attention given to the fifth assessment results, except with regard to trend information or other special considerations.

The purpose of this introductory chapter is to provide the opportunity for readers to learn about NAEP. That is, the chapter will set the context for the chapters that follow by summarizing relevant information about the sixth mathematics assessment. For information on the fifth and sixth assessments, the reader should consult relevant sections of *The NAEP Guide: A Description of the Content and Methods of the 1990 and 1992 Assessments* (Mullis 1991). Chapter 2 in this volume provides additional information on the fifth and sixth mathematics assessment.

FRAMEWORK AND OBJECTIVES

Because of NAEP's conscious effort to be sensitive to changes in curriculum and educational objectives, some changes are made in the assessment each time a curriculum subject is measured. Like their predecessors, the fifth and sixth NAEP mathematics assessments were organized around a set of mathematical processes and content-area objectives, referred to as a "framework." This, in turn, guided the development and selection of the test questions, called "items" or "cognitive items" by NAEP. As a result of the authorization of state-level NAEP, the framework for the fifth and sixth assessments was newly developed in 1988 through a consensus process involving groups of mathematics educators, policymakers, and mathematics supervisors from all fifty states.

The NAEP mathematics frameworks have evolved over time to reflect the "best thinking" of the mathematics education community with respect to mathematics content and mathematical processes. At the time that the framework for the fifth and sixth assessments was under development, the best thinking was guided by an early draft of the NCTM *Curriculum and Evaluation Standards for School Mathematics* (1989), a document that reflects a vision of what it means to be mathematically literate in today's technological society. The document focuses heavily on problem solving, communication of mathematical ideas, critical reasoning, and connections to ideas and procedures both within mathematics and across other content areas. The Mathematics Objectives Committee, the group responsible for development of the NAEP mathematics framework, drew upon that early draft of the NCTM *Standards* during its deliberations (National Assessment of Educational Progress 1988) and thereby demonstrated the potential for the alignment of the NAEP assessment with the vision for the school mathematics curriculum in this country. Additional information about the development of the framework for the fifth and sixth NAEP mathematics assessments can be found in chapter 2.

The framework itself was organized as a matrix with fifteen cells involving five content areas (Numbers and Operations; Measurement; Geometry; Data Analysis, Statistics, and Probability; Algebra and Functions) and three ability categories (Conceptual Understanding; Procedural Knowledge; Problem Solving). The matrix appears in figure 1.1. Brief descriptions of the content areas and ability categories appear in the sections that follow. Additional information can be found in the framework document itself (NAEP 1988).

CONTENT / ABILITY	Numbers and Operations	Measurement	Geometry	Data Analysis, Statistics, and Probability	Algebra and Functions
Conceptual Understanding					
Prodecural Knowledge					
Problem Solving					

Fig.1.1 Framework for the sixth assessment
Source: National Assessment of Educational Progress, *Mathematics Objectives: 1990 Assessment,* 1988

Content Areas

The five content areas assessed by NAEP were selected to represent topics commonly studied by the majority of students at a given grade level. The percent distribution of items by content area suggested in the mathematics framework document appears in table 1.1. The distribution of items by content was established in the interest of creating a more forward-thinking assessment that gave greater emphasis to content areas such as Geometry and Algebra and Functions and less to Numbers and Operations than in the past.

Each of the content categories as defined by NAEP is described next. Additional information about each content area is provided in the chapter that discusses and interprets results from that area. (For example, chapter 6 contains information about the content area of Measurement in the NAEP framework.)

Numbers and Operations. This content area focuses on students' understanding of whole numbers, fractions, decimals, and integers and their

Table 1.1 Percent Distribution of Items by Grade and Content Area

	Grade 4	Grade 8	Grade 12
Numbers and Operations	45	30	25
Measurement	20	15	15
Geometry	15	20	20
Data Analysis, Statistics, and Probability	10	15	15
Algebra and Functions	10	20	25

Source: National Assessment of Educational Progress, *Mathematics Objectives: 1990 Assessment*, 1988

application to computational, estimation, and real-world situations. Understanding numerical relationships as expressed in ratios, proportions, and percents is also included.

Measurement. This content area focuses on students' ability to describe real-world objects using numbers. Important abilities assessed here include identification of attributes, selection of appropriate units, application of measurement concepts such as perimeter and area, and communication of measurement-related ideas in writing.

Geometry. This content area focuses on students' knowledge of geometric figures in one, two, and three dimensions and their ability to work with and explain geometric relationships. The majority of topics in this content area, especially those topics appropriate for fourth- and eighth-grade students, involve informal geometry and spatial sense. Topics appropriate for twelfth-grade students include coordinate geometry and vectors.

Data Analysis, Statistics, and Probability. This content area focuses on data representation and analysis in real-world situations. Topics include appropriate methods for gathering data, the visual exploration of data, and the development and evaluation of arguments based on data analysis.

Algebra and Functions. Of the five content categories, this category is the broadest in scope, covering a significant portion of the topics in the grades 9–12 curriculum such as algebra, elementary functions, trigonometry, and discrete mathematics. Algebraic and functional topics appropriate for elementary and middle school students are treated in informal, exploratory ways. Proficiency in this content area involves the ability to use algebra as a means of representation and to use algebraic skills and concepts as problem-solving tools.

Ability Categories

The three categories of mathematical abilities were selected to encourage the development of a broad range of test items that measured different levels of thinking. Unlike the distribution of items by content area in which the percents differed between and across grade levels, the distribution of items by grade and ability category was the same for grades 4, 8, and 12: 40 percent of the items were classified as Conceptual Understanding, 30 percent as Procedural Knowledge, and 30 percent as Problem Solving. Each category as defined by NAEP is described next.

Conceptual Understanding. Conceptual understanding is essential to performing procedures in a meaningful way and has an important connection to problem solving. Students demonstrate conceptual understanding in a variety of ways, including recognizing, labeling, and generating examples and counterexamples; using models, diagrams, and symbols; identifying and using principles; knowing and applying facts and definitions; making connections among different modes of representations; comparing, contrasting, and integrating concepts; recognizing, interpreting, and applying symbols to represent concepts; and interpreting assumptions and relations.

Procedural Knowledge. Procedural knowledge includes the various numerical algorithms in mathematics that have been created as tools to meet specific needs in an efficient manner. In NAEP, reading and producing graphs and tables, executing geometric constructions, and performing noncomputational skills such as rounding and ordering are also considered "procedures." Students demonstrate procedural knowledge by providing evidence of their ability to select and apply appropriate procedures correctly, by verifying and justifying the correctness of a procedure, and by extending or modifying procedures.

Problem Solving. Problem solving includes the ability to recognize and formulate problems; determine the sufficiency and consistency of data; use strategies, data, models, and relevant mathematics; use reasoning (i.e., spatial, inductive, deductive, statistical, proportional); and judge the reasonableness and correctness of solutions. An important characteristic of problem solving in NAEP is that students are required to use their reasoning and analytic skills in the context of new situations.

The Validity of the Framework and Objectives

Through a series of independent, congressionally mandated studies concerning the validity of the 1990 and 1992 NAEP tests used in the state-level assessments, representatives of the mathematics education community had the opportunity to evaluate the framework and objectives. Findings from those studies related to the mathematics assessments (Silver and Kenney

1994; Silver, Kenney, and Salmon-Cox 1992) reveal that expert panels composed of classroom teachers and teacher educators agreed that the framework was reasonable and represented an attempt to balance the vision represented in the newly published NCTM *Curriculum and Evaluation Standards* with the reality of instructional practice in mathematics in the early 1990s. Despite the view that the framework was reasonable, there was considerable concern about the limitations imposed by a "matrix view" of mathematics assessment that required each item to be classified in only one content area and only one ability category.

One finding from the expert-panel studies had a direct bearing on the structure of this volume of interpretive reports. When asked to classify items according to the content and ability categories in the framework, panelists were more likely to agree with the content classification than with the ability classification. This result was not unexpected; even the panels that developed the framework acknowledged that the ability categories were "not discrete or mutually exclusive, but rather highly integrated" (NAEP 1988, p. 11). Given the acknowledgment of the framework-development panel and the findings from the external studies and given the precedent of using NAEP content areas instead of process or ability categories as the basis for chapters in past interpretive reports published by NCTM, we chose to continue the pattern of interpreting performance results according to the five content areas in the framework.

COGNITIVE ITEMS

Frameworks provide the template for the assessment design and in particular for the development of the cognitive items appearing on the tests. Using the framework just described, a group of mathematics educators worked with the staff of the Educational Testing Service (ETS), the NAEP contractor, to develop the items for the sixth assessment. The process used to develop the items was similar to that described by Carpenter (1989) for the fourth assessment. Items on the sixth assessment appeared in either multiple-choice or constructed-response formats, and students were provided with, and permitted to use, mathematical tools such as rulers and calculators while working on certain sets of items.

Item Formats

The sixth assessment consisted of three formats for items: multiple-choice, regular constructed-response, and extended constructed-response. Examples of each item format appear in the content-based chapters (chapters 5–9) in this volume. About 70 percent of the items on each grade-level test were multiple-choice items and about 30 percent were regular or extended constructed-response items.

Multiple-choice and regular constructed-response items were included on the fourth assessment as well as on previous assessments. In the sixth assessment, the multiple-choice items administered to fourth-grade students had four choices, and those administered to students in grades 8 and 12 had five choices. Regular constructed-response items were open-ended questions that asked students to formulate their own numerical answers or to write a brief explanation or justification for a mathematical situation. (An example of a regular constructed-response item appears in chapter 8, table 8.3.) Students' answers to regular constructed-response items were scored by trained raters according to a right-or-wrong scoring scheme.

The extended constructed-response questions were pilot-tested on the fifth assessment and became a regular feature on the sixth assessment. These tasks not only required students to construct their own responses but also provided students with an opportunity to express their mathematical ideas in writing and to demonstrate their depth of understanding. The fourth-grade test included five extended tasks, and the eighth- and twelfth-grade tests each included six extended tasks. (Examples of extended questions at each grade level appear in these chapters: chapter 5, table 5.15 [grade 4 question]; chapter 9, table 9.5 [grade 8 question]; and table 9.13 [grade 12 question].) Responses to these extended tasks were scored by trained raters using a partial-credit, focused holistic scoring scheme based on the categories of minimal, partial, satisfactory, and extended.

Detailed information about student performance on some regular and all extended constructed-response items that were released to the public was included in an official NAEP report (Dossey, Mullis, and Jones 1993). The content-based chapters in this volume have made use of information showcased in that report. Because we had the opportunity to gather samples of responses to selected regular and extended questions and to analyze them qualitatively, the content chapters also contain new information about performance on the released regular and extended constructed-response questions not available from the Dossey, Mullis, and Jones report.

Despite NAEP's continuing attempt to improve the items on the grade-level tests, the items from the sixth assessment have some limitations. In particular, because of the forced-choice format of multiple-choice items and the guessing factor inherent in them, it is difficult to identify with certainty the reasons why students selected any particular choice. Authors of the content-based chapters, while acknowledging the guessing factor, have chosen to speculate on misconceptions and error patterns on the basis of performance on multiple-choice items and have tried to link their inferences to relevant literature in the field of mathematics education.

Because of our attempts to gather and then analyze student responses to some released regular constructed-response questions and to all released extended questions, the authors were able to expand the limited information available from official NAEP sources about student performance on those questions. In particular, examining student responses enabled the authors to make stronger inferences about students' strengths and weaknesses in particular areas such as reasoning about numbers, drawing geometric figures, interpreting pictographs, recognizing patterns, and reasoning algebraically. Again, where possible, references to mathematics education research were included to lend support to inferences made by the authors.

Mathematical Tools

While working on some items, students were permitted to use tools such as rulers, protractors, manipulatives in the form of geometric shapes, and calculators. However, use of these tools was restricted to particular sets of items (called "item blocks" by NAEP); students did not have access to these tools at all times during the assessment.

The use of rulers and protractors has been included on previous NAEP assessments. Items involving the ruler or protractor required students to make measurements (for example, determine the degree measure of a given angle) or to provide a fairly accurate representation of a geometric shape (for example, drawing a rectangle of given proportions). The questions involving the use of manipulatives in the form of geometric shapes were new to the sixth assessment, and they asked students to assemble the shapes to create new shapes having particular properties or areas.

Calculators have been included in the NAEP mathematics assessment since the second assessment administered in 1978. For the fourth assessment students' ability to use calculators was assessed through the use of "calculator blocks"; that is, on certain blocks of items, students were given a simple, four-function calculator, instructed in its use, and then permitted to use the calculator while working on computation and routine application items. Many of the same items in the calculator blocks also appeared in other blocks on which students were *not* permitted to use calculators. This structure permitted a comparison of performance with and without calculators. (See chapter 8 in Lindquist [1989] for a discussion of those results.)

The sixth assessment also was structured so that students had calculators for particular blocks of items. While working on the items in a calculator block, students in grade 4 were supplied with a Texas Instruments TI-108 calculator (a simple four-function calculator), and students in grades 8 and 12 received a TI-30 Challenger (a scientific calculator). Only while working on items in a calculator block were students permitted to use the calculator, and they were provided with a short introduction to the use of the calculators before beginning the assessment exercises (Mullis 1991). Further instructions on calculator use were printed on the page preceding the

items in a calculator block, and a reference card accompanied the calculators given to eighth- and twelfth-grade students.

Items in the calculator blocks were classified in three ways—calculator-
inactive items, calculator-active items, and calculator-neutral items—
according to the definitions provided in the framework document. The definitions appear in the framework document (NAEP 1988, p. 33) as follows:

> Calculator-inactive items are those whose solution neither requires nor sug-
> gests the use of a calculator; in fact, a calculator would be virtually useless as an
> aid to solving the problem. Calculator-neutral items are those in which the so-
> lution to the question does not require the use of a calculator. Given the option,
> however, some students might choose to use a calculator to perform numerical
> operations. In contrast, items classified as "calculator active" require calculator
> use; a student would likely find it almost impossible to solve the question with-
> out the aid of a calculator.

In the sixth assessment, each calculator block contained all three kinds
of calculator-block items, but calculator-block items did not appear in any
other item block within a grade level. This structure stands in contrast to
that used in the fourth assessment, which ensured that some items were
administered twice: both in a designated calculator-block *and* a block for
which students were not permitted to use calculators. Thus, for the sixth
assessment it was not possible to compare item-by-item performance for
students who had access to calculators and students who had no such access. Providing some information about calculator use and performance
on calculator-block items using 1992 results required a different kind of
analysis. To find patterns in performance on calculator-block items and
reported calculator use, data were obtained that linked percent-correct values on calculator-block items to the yes-or-no responses to the calculator-
use self-report question. Where relevant, these results are reported within
the content-based chapters.

Although NAEP's commitment to allowing students to use various
mathematical tools while working on some items is laudable, there are some
important concerns associated with the way those tools have been incor-
porated into the assessment, the most obvious being that the tools were
permitted to be used only with particular blocks of items and were not
available for others. For the calculators in particular, while there is some
political value in permitting students to use calculators on a national-level
assessment of mathematics achievement and while providing each student
with the same type of calculator addresses equity concerns that have arisen
during debates on the role of calculators in assessment (e.g., Wilson and
Kilpatrick 1989), the restricted use of calculators on NAEP represents an
artificial situation, inconsistent with the view espoused in the NCTM *Curriculum and Evaluation Standards* that "calculators should be available for
all students at all times" (1989, p. 8).

QUESTIONNAIRES

In addition to cognitive items, the sixth assessment included questionnaires administered to students, teachers, and principals. Each questionnaire was developed according to relevant categories in five broad educational areas: instructional content, instructional practices and experiences, teacher characteristics, school conditions and context, and conditions beyond school (i.e., home support, out-of-school activities, and attitudes). Results from these three questionnaires were used in nearly all chapters in the volume. In particular, chapter 3 on race/ethnicity and gender and chapter 4 on characteristics of mathematics teachers make extensive use of the questionnaire data.

Student Questionnaire

Each student in the grade-level NAEP samples completed a three-part questionnaire. The first two parts were administered before the cognitive-item sections. The first part of the student questionnaire included demographic questions about race/ethnicity, language spoken in the home, mother's and father's level of education, reading materials in the home, homework, attendance, school climate, academic expectations, and which parents live at home. The second part included questions about educational experiences in mathematics such as instructional activities, courses taken, use of specialized resources such as calculators and computers, and views about the utility and value of mathematics. The third part of the student questionnaire was administered after a student completed the cognitive items and contained questions about students' familiarity with the format of the NAEP items and whether they tried to do their best work.

Teacher Questionnaire

The teacher questionnaire was administered to all mathematics teachers of fourth- and eighth-grade students participating in the mathematics assessment. Since relatively few twelfth-grade students who participated in the assessment were enrolled in mathematics classes, there were no NAEP teacher questionnaires given to twelfth-grade teachers.

The questionnaire contained two parts. The first part pertained to the teachers' background and training and contained questions about gender, race, ethnicity, years of teaching experience, certification, academic degrees, major areas of study, in-service training, and others. The second part of the questionnaire pertained to the procedures a teacher used for each class containing a student assessed by NAEP and included questions about the ability level of students in the class, whether students were assigned to the class by ability level, homework assignments, frequency of instructional activities used in class, opportunity to learn, and use of particular resources such as calculators, manipulatives, and computers.

School Characteristics and Policies Questionnaire

This questionnaire was given to the principal (or other administrator) of each school that participated in NAEP. The questionnaire collected information about background and characteristics of school principals, length of school day and year, school enrollment, absenteeism, dropout rates, size and composition of teaching staff, policies about tracking, curriculum, testing practices and use, special services, policies for parental involvement, and schoolwide problems.

SAMPLES OF STUDENTS AND ADMINISTRATION PROCEDURES

Because Educational Testing Service was retained as the NAEP contractor for the fifth and sixth assessments, the sampling and administration procedures used for those assessments were similar to those developed for the fourth assessment and described by Carpenter (1989). The next sections describe sampling and administration procedures unique to the sixth assessment. Again, the reader can find more detailed information about sampling and administration procedures in Mullis (1991).

Students taking the 1992 assessment were selected according to a complex procedure that identified a representative national sample for each of grades 4, 8, and 12. All together, 8,738 fourth-grade students, 9,432 eighth-grade students, and 8,499 twelfth-grade students participated in the assessment, which was administered during the months of February–April 1992. In previous assessments, students were selected by age, and in the fourth assessment they were selected at grades 3, 7, and 11, so the results for the fifth and sixth assessments provide a somewhat different perspective on performance than previous assessments at the national level.

In the NAEP design, no student takes all items at a particular grade level. Instead, the items were divided into blocks of items, that is, sets of items that always remain together. In 1992, each grade-level test had thirteen blocks of items, with the number of items in each block varying from about ten to twenty-five items. The thirteen blocks of items were assembled into twenty-six booklets; each booklet contained the student questionnaire, three blocks of items, and the motivational questionnaire as well as appropriate instructions.

Each student who participated in the assessment took only the mathematics assessment and completed the questionnaires and items in only one booklet. Students were allowed fifteen minutes for each item block, and they worked at their own pace through each block. Because "speededness" was identified as a problem in the fourth assessment (see Carpenter [1989]), the number of items in a block was reduced for the fifth and sixth assessments in an attempt to lessen the nonresponse rates for contiguous items appearing at the end of an item block. For the fourth assessment, Carpenter (1989) reported that the nonresponse rate (called "not reached" rate by NAEP) for some items at the end of their item blocks was as high as 50 percent for a number of items, which resulted in a less accurate estimate of performance on those items for the population of students as a whole. For the sixth assessment, however, the not-reached rates were substantially lower than those for the fourth assessment, with most not-reached rates for the last two items in a block at 20 percent or less. We report not-reached rates where relevant to the discussion of particular items in the content chapters.

REPORTING AND INTERPRETING RESULTS

One of the most dramatic changes between the fourth and fifth and the sixth mathematics assessments involved the primary method used to report student achievement results. In official NAEP reports for the fourth and fifth assessments, the primary reporting method used a standard score scale ranging from 0 to 500 with a mean of 250. To enhance the interpretability of scores, descriptions were provided about the kinds of mathematics that students know and can do at major points on the scale, called "anchor points." These anchor points were at score levels 200, 250, 300, and 350 but contained no grade-level information. (For information on the scale-anchoring process used for the 1990 assessment, the reader should consult Mullis et al. [1991].) However, in 1992 reporting by "achievement levels" became the primary reporting method for the sixth assessment. The achievement levels, designated as *Basic, Proficient,* and *Advanced,* were set for each grade level and described how students *should* perform on the content areas in the assessment. (For information on the achievement levels used for the 1992 assessment, the reader should consult Mullis et al. [1993].) The method of reporting NAEP results by achievement levels has not been without controversy and has been scrutinized and criticized in a number of reports (for example, U.S. General Accounting Office [1993]; National Academy of Education [1993]; Silver and Kenney [1993]). In particular, Silver and Kenney identified an important source of the problem with the achievement levels established for the sixth assessment: the retrofitting of the levels to a test that had not been designed to be reported with respect to definitions of Basic, Proficient, and Advanced performance at grades 4, 8, and 12.

Previous interpretive reports of NAEP mathematics results have found the concept of a total score to have little meaning, especially in the case of interpreting performance in content areas. Instead, writers of those reports preferred to examine results on the basis of percent-correct values for clusters of related items and percent-responding values for the choices or score categories on individual items and then to interpret those results in light of current curriculum and classroom practices. Beginning with the fourth assessment, some chapters in the interpretive reports augmented item-level percent-correct values with other reporting methods such as scale scores and scale-anchor descriptions, especially when it was deemed important to report overall performance (Dossey et al. 1989), performance by race/ethnicity subgroups groups (Johnson 1989), and performance by gender (Meyer 1989).

In this report for the sixth assessment, chapter authors have used NAEP results reported in a variety of ways, focusing on reporting methods that were deemed the most appropriate within each chapter. For example, the chapters concerned with overall performance by selected subgroups (chapter 3) and by grade levels according to teacher characteristics (chapter 4) make use of scale scores or achievement levels. As was the case for the content-based chapters in the previous interpretive report of NAEP results, the content-based chapters in this volume (chapters 5–9) are based primarily on item-level data such as percent-correct values. However, scale scores also appear in each content chapter to provide a glimpse of student performance in that content area across grade levels and according to subgroups of students.

CONCLUSION

Learning about NAEP can be daunting: NAEP is one of the most complex assessments ever designed and administered on a large-scale basis. This introductory chapter has attempted to provide information about the NAEP mathematics assessment so that the reader can better understand NAEP and the interpretations offered by the writers in this volume. To assist interested readers in learning more about NAEP and about the various reports produced by government sources or through the efforts of NCTM, a bibliography of reports, journal articles, and book chapters is included at the end of this volume.

In the current climate of educational reform, NCTM has emerged as a leader in establishing curriculum standards, teaching standards, and assessment standards for school mathematics, and it has consistently looked to NAEP results as one indicator of what American students know and can do in mathematics. It is hoped that continuing the tradition of the NCTM interpretive reports of NAEP results will have an impact on the mathematics education community. In particular, for classroom teachers and teacher educators, the in-depth interpretive look at the results from the sixth NAEP contained in this volume will provide valuable information about student strengths in mathematics and areas that need additional educational attention.

REFERENCES

Carpenter, Thomas P. "Introduction." In *Results from the Fourth Mathematics Assessment of the National Assessment of Educational Progress*, edited by Mary Montgomery Lindquist, pp. 1–9. Reston, Va.: National Council of Teachers of Mathematics, 1989.

Dossey, John A., Ina V. S. Mullis, and Chancey O. Jones. *Can Students Do Mathematical Problem Solving? Results from Constructed-Response Questions in NAEP's 1992 Mathematics Assessment.* 23-FR-01. Washington, D.C.: National Center for Education Statistics, August 1993.

Dossey, John A., Ina V. S. Mullis, Mary M. Lindquist, and Donald L. Chambers. "What Can Students Do? (Level of Mathematics Proficiency for the Nation and Demographic Subgroups)." In *Results from the Fourth Mathematics Assessment of the National Assessment of Educational Progress*, edited by Mary Montgomery Lindquist, pp. 117–34. Reston, Va.: National Council of Teachers of Mathematics, 1989.

Johnson, Martin L. "Minority Differences in Mathematics." In *Results from the Fourth Mathematics Assessment of the National Assessment of Educational Progress*, edited by Mary Montgomery Lindquist, pp. 135–48. Reston, Va.: National Council of Teachers of Mathematics, 1989.

Lindquist, Mary Montgomery, ed. *Results from the Fourth Mathematics Assessment of the National Assessment of Educational Progress*. Reston, Va.: National Council of Teachers of Mathematics, 1989.

Meyer, Margaret R. "Gender Differences in Mathematics." In *Results from the Fourth Mathematics Assessment of the National Assessment of Educational Progress*, edited by Mary Montgomery Lindquist, pp. 149–59. Reston, Va.: National Council of Teachers of Mathematics, 1989.

Mullis, Ina V. S. *The NAEP Guide: A Description of the Content and Methods of the 1990 and 1992 Assessments.* Princeton, N.J.: Educational Testing Service, 1991.

Mullis, Ina V. S., John A. Dossey, Eugene H. Owen, and Gary W. Phillips. *NAEP 1992 Mathematics Report Card for the Nation and the States: Data from the National and Trial State Assessments.* 23-ST-02. Washington, D.C.: National Center for Education Statistics, April 1993.

_____. *The STATE of Mathematics Achievement: NAEP's 1990 Assessment of the Nation and the Trial Assessment of the States.* 21-ST-04. Washington, D.C.: National Center for Education Statistics, June 1991.

National Academy of Education. *Setting Performance Standards for Student Achievement.* Stanford, Calif.: National Academy of Education, 1993.

National Assessment of Educational Progress. *Mathematics Objectives: 1990 Assessment.* Princeton, N.J.: Educational Testing Service, National Assessment of Educational Progress, 1988.

National Council of Teachers of Mathematics. *Curriculum and Evaluation Standards for School Mathematics.* Reston, Va.: National Council of Teachers of Mathematics, 1989.

Silver, Edward A., and Patricia Ann Kenney. "The Content and Curricular Validity of the 1992 NAEP TSA in Mathematics." In *The Trial State Assessment: Prospects and Realities: Background Studies,* pp. 231–84. Stanford, Calif.: National Academy of Education, 1994.

_____. "Expert Panel Review of the 1992 NAEP Mathematics Achievement Levels." In *Setting Performance Standards for Student Achievement: Background Studies,* pp. 215–81. Stanford, Calif.: National Academy of Education, 1993.

Silver, Edward A., Patricia Ann Kenney, and Leslie Salmon-Cox. "The Content and Curricular Validity of the 1990 NAEP Mathematics Items: A Retrospective Analysis." In *Assessing Student Achievement in the States: Background Studies,* pp. 157–218. Stanford, Calif.: National Academy of Education, 1992.

U.S. General Accounting Office. *Educational Achievement Standards: NAGB's Approach Yields Misleading Interpretations.* Report no. GAO/PEMD-93-12. Washington, D.C.: General Accounting Office, 1993.

Wilson, James W., and Jeremy Kilpatrick. "Theoretical Issues in the Development of Calculator-Based Mathematics Tests." In *The Use of Calculators in the Standardized Testing of Mathematics,* edited by John Kenelly, pp. 7–15. New York: College Entrance Examination Board, 1989.

2

NAEP Mathematics—1990–1992: The National, Trial State, and Trend Assessments

John A. Dossey & Ina V. S. Mullis

THE MATHEMATICS assessments conducted by NAEP serve as one of the barometers of progress toward meeting the NCTM *Standards* for curriculum, teaching, and assessment (NCTM 1989; 1991; 1995). In doing so, these assessments provide a wealth of information concerning the status of mathematics teaching and learning in our schools at grades 4, 8, and 12.

The NAEP tests are developed to measure the condition of student achievement in mathematics and to monitor trends across time. Through questionnaire data, NAEP provides a picture of the mathematics teaching force, students' educational backgrounds in mathematics, and the instructional methods used in classrooms. To maintain relevant baseline data to measure future progress, the mathematics assessments are constantly updated to reflect the current nature of the curriculum. An effort is made to reflect the essence of new initiatives in teaching and assessments at their onset in order to provide pertinent measures of progress in achievement. Although the NAEP assessments attempt to evolve and to keep abreast of current methods, they measure what does take place in the schools rather than what might take place in the schools. This means trying to keep up with current reforms as much as is feasible but not leading the vanguard. As such, some consider the NAEP tests to be too conservative in their composition and approach. Maintaining the stability required for measuring trends, while still introducing innovations, has led to complexities in NAEP. NAEP has provided for links to the future and links to the past by conducting separate mathematics assessments for different purposes (Mullis 1991). The kinds of NAEP assessments are described next.

THE NAEP ASSESSMENTS

There are three kinds of NAEP assessments: national, state-level, and trend. The national mathematics assessments have been given for more than twenty years on a quasi-regular basis—1973, 1978, 1982, 1986, 1990, and 1992. Substantial portions, but not all parts, of these assessments incorporate newer, more innovative methods. Generally, however, to ensure that the trend results reflect changes in student performance and not changes in the test, many of the questions—and most of the procedures—are held constant from assessment to assessment. An exception was the 1990 mathematics assessment, which was entirely new and developed in concert with the initial implementation of state-level assessments. To ensure that participants in NAEP are representative of students across the nation, they are selected on the basis of a stratified, three-stage sampling plan. First, counties are classified by region and community type and then randomly selected. Second, schools—both public and private—are classified and randomly selected. Finally, students are randomly selected within schools.

In 1990, NAEP also began providing mathematics data at the state level. First offered on a trial basis, the state-level assessments were given at grade 8 in 1990 and at grades 4 and 8 in 1992. This voluntary program for states replicated most of the questions included in the 1990 and 1992 national assessments, thus providing comparative data to the nation as well as among states. The basic sampling design for the state assessments was to select a sample of 100 public schools from each state, with a sample of thirty students drawn from each school. In 1992, representative samples of fourth- and eighth-grade students attending public schools were assessed in each of forty-four states and territories. Comparable trend results are available for the nation and for the thirty-seven states and territories that participated in the 1990 and 1992 programs at grade 8.

In addition to national and state-level assessments, NAEP has also administered a trend assessment at the national level. In the basic subject areas, including mathematics, NAEP assesses nationally representative samples of students by using the methods of past assessments. These assessments are used to maintain trends established across 20 years. Initially administered as part of the national assessments, the trend assessments have been kept constant since 1986, when the two types of assessments became separate. The trend assessments sample students at ages 9, 13, and 17, which is consistent with the procedures NAEP used until the mid-1980s. Also, the trend assessments reflect the 1970s curricular emphasis on basic skills. Even though it is important for students to continue improving in the fundamentals, recent national assessments are more appropriate for monitoring growth in mathematics problem solving.

In the remainder of this chapter, we examine the global picture of the status of mathematics education and changes in this picture revealed by these various forms of NAEP assessments. The focus is on results from the 1990 and 1992 assessments at the national and state levels as well as results from the NAEP trend assessments in mathematics.

THE NAEP NATIONAL ASSESSMENT

When, in 1988, Congress added a new dimension to NAEP by authorizing the first state-level assessments in 1990 and 1992, NAEP began playing an increasingly visible role in measuring student achievement. The trial state assessment program provided an opportunity to start over, and NAEP attempted to initiate new innovative systems. To this end, the Council of Chief State School Officers was awarded a special National Assessment Planning Project to develop a framework for the 1990 assessment that would reflect the current status of mathematics teaching and learning in grades 4, 8, and 12 and, at the same time, provide a basis for examining the impact of the NCTM *Curriculum and Evaluation Standards* on school mathematics.

NAEP's 1990 and 1992 mathematics assessment framework specified five content areas (Numbers and Operations; Measurement; Geometry; Data Analysis, Statistics, and Probability; Algebra and Functions) and three process areas (Conceptual Understanding, Procedural Knowledge, Problem Solving) (National Assessment of Educational Progress 1988). The project to develop the framework involved widespread participation and review, including state committees of mathematics educators, policymakers, practitioners, and citizens at large.

The newly developed 1990 mathematics assessment featured more questions requiring students to construct their own answers and included the use of scientific calculators. Estimation and complex problem solving were assessed in a special study using audiotapes that paced students through the questions. In 1992, an even greater focus was placed on open-ended problem solving, including some tasks in which students were given rulers or geometric shapes to use in formulating their responses.

Student responses to the items on the national NAEP mathematics assessment were then scaled and placed on a proficiency scale of 0 to 500. This scale represented the full range of student proficiency that might be expected across the grades. While such a scale provides a picture of student progress relative to performance across grade levels and between and among demographic subpopulations, it does not describe what students know and are able to do in mathematics nor does it evaluate students' performance against a nationally held standard of performance concerning what students should be able to do in mathematics.

To accomplish this, the National Assessment Governing Board (NAGB), the group authorized by Congress to formulate policy for NAEP, established three levels of achievement to describe student performance in mathematics: *Basic, Proficient,* and *Advanced* (NAGB 1990; Mullis et al. 1993). These achievement levels are defined in figure 2.1. Performance at the Basic level denotes partial mastery of the knowledge, content, and skills for proficient work at each grade level, but not work that is deemed

Basic   This level, below proficient, denotes partial mastery of knowledge and skills that are fundamental for proficient work at each grade— 4, 8, and 12. For 12th grade, this is higher than minimum competency skills (which are normally taught in elementary and junior high schools) and covers significant elements of standard high-school-level work.

Proficient The central level represents solid academic performance for each grade tested—4, 8, and 12. It reflects a consensus that students reaching this level have demonstrated competency over challenging subject matter and are well prepared for the next level of schooling. At grade 12, the proficient level encompasses a body of subject-matter knowledge and analytical skills, of cultural literacy and insight, that all high school graduates should have for democratic citizenship, responsible adulthood, and productive work.

Advanced This highest level signifies superior performance beyond proficient grade-level mastery at grades 4, 8, and 12. For 12th grade, the advanced level shows readiness for rigorous college courses, advanced technical training, or employment requiring advanced academic achievement.

Fig. 2.1 Definitions of the achievement levels
Source: Mullis et al. *NAEP 1992 Mathematics Report Card for the Nation and the States,* 1993

satisfactory at that level. Proficient performance represents solid academic performance at each grade level. Achievement at the Advanced level signifies superior performance at each of the grades. The achievement levels were first developed for use with the national NAEP in 1990. On the basis of feedback and analysis of the relationship of these levels to the traditional 0 to 500 mathematical proficiency scale, the definitions of the achievement levels were readjusted, which resulted in the form used to report NAEP results in 1992 (Phillips et al. 1993).

Students' Mathematics Achievement

The results from the 1990 and 1992 national NAEP assessments evoke both a sense of despair and hope (Mullis et al. 1993). The overall levels of student performance at each of the grades assessed are considerably lower than what mathematics educators would hope to find. Table 2.1 contains information about the percent of students performing at or above each of the three achievement levels—Basic, Proficient, and Advanced—and at a level that is "Below Basic" for 1990 and 1992. The data in the table show that students in all three grade levels made significant gains in reaching the achievement levels between 1990 and 1992. In particular, the data indicate a larger increase in overall proficiency at grade 4 but stronger gains at the upper levels at grade 8. When comparisons of gains are limited to the advanced level, there were small gains at grades 4 and 8, but none at grade 12.

The results of NAEP's 1992 mathematics assessment indicate that student performance is improving nationally but that a considerable challenge remains. Proportions of students at the higher achievement levels continue to be low, particularly for those subpopulations of students historically considered to be "at risk." For the nation, there were statistically significant increases in average mathematics proficiency between 1990 and 1992 for fourth-, eighth-, and twelfth-grade students. Despite these positive findings, slightly more than 60 percent of the students in grades 4, 8, and 12

Table 2.1 National Percents of Students according to Mathematics Achievement Levels, Grades 4, 8, and 12

	1990	1992
Grade 4		
Percent below Basic	46(1.4)	39(1.0)<
Percent at or above		
Basic	54(1.4)	61(1.0)>
Proficient	13(1.1)	18(1.0)>
Advanced	1(0.4)	2(0.3)
Grade 8		
Percent below Basic	42(1.4)	37(1.1)<
Percent at or above		
Basic	58(1.4)	63(1.1)>
Proficient	20(1.1)	25(1.0)>
Advanced	2(0.4)	4(0.4)
Grade 12		
Percent below Basic	41(1.5)	36(1.2)<
Percent at or above		
Basic	59(1.5)	64(1.2)>
Proficient	13(1.0)	16(0.9)
Advanced	2(0.3)	2(0.3)

>The value for 1992 was significantly higher than the value for 1990 at about the 95 percent confidence level.

<The value for 1992 was significantly lower than the value for 1990 at about the 95 percent confidence level.

Note: The standard errors of the estimated percents appear in parentheses.

Source: Mullis et al. *NAEP 1992 Mathematics Report Card for the Nation and the States,* 1993

were estimated to be at or above the Basic level on the 1992 mathematics assessment. At this level, students should exhibit partial mastery of the knowledge and skills fundamental for proficient work. Nationally, across the three grades, 25 percent or fewer students were estimated to be at the Proficient level or beyond, where students should exhibit evidence of solid academic performance. The percents of students attaining the Advanced level, where students should exhibit superior performance, ranged from an estimated 2 to 4 percent. At best, the NAEP data indicate that we are making progress in a situation where achievement still falls far short of that expected for students in U.S. classrooms.

Results for Subgroups of Students

The sampling design for the national NAEP allows for examining a number of student demographic variables and contexts for learning mathematics. In both 1990 and 1992, two-thirds or more of the Asian/Pacific Islander and White students were estimated to have achievement at or above the Basic level, while fewer than one-half of the American Indian, Black, and Hispanic students demonstrated achievement at this partial mastery level. Gender differences were not large. There were no significant differences at grades 4 and 8, but males tended to outperform female students at grade 12. For example, 18 percent of the males at this level were estimated to be at or above the Proficient level compared to 14 percent of the females.

Students attending schools in advantaged urban areas had the largest gains in proportions of students reaching the Proficient or Advanced levels of mathematics achievement; their counterparts in disadvantaged urban areas fared worst of all. However, these gains from 1990 to 1992 were significant only at grade 8. Another comparison was made between the schools ranking in the top one-third of those participating in the NAEP sample and those in the bottom third. The analysis of the gains made between these groups shows that students in the schools ranking in the top one-third performing schools at both grades 4 and 8 placed more students in higher achievement groups in 1992 than the comparable group did in the 1990 assessment. Similarly, the higher the educational level of a student's better-educated parent, the more likely the student was to be included in higher achievement groups.

One of the most interesting findings from the national NAEP in 1992 is the comparative analysis of student performance for students attending public, Catholic, and other private schools. While no significant differences were found between the three types of schools in average proficiency at the three grades, greater percentages of students attending Catholic or other private schools tended to perform in the upper levels of the proficiency scales.

NAEP's Focus on Problem Solving

Perhaps the most revealing aspect of the 1992 national NAEP was the focus given to items requiring students to construct their own answers (Dossey, Mullis, and Jones 1993). While NAEP had long required students to provide answers to some items on the tests, most of these items went little beyond multiple-choice questions without the choices. The 1992 NAEP moved to asking students to both answer questions and provide rationales for their responses. These questions took two forms. The first of these items were called regular constructed-response items. These items were graded on a right-wrong basis, but with scoring done according to set criteria for what constituted evidence for awarding credit. An example of such a question is the Trash Cans problem discussed in chapter 8.

Smaller percentages of eighth-grade students were able to answer correctly such questions than they were comparable multiple-choice questions in the same content areas. The inability of students to perform in this format was even more clearly underscored by their achievement on extended constructed-response items. These items required students to write an extended response to answer a question or series of questions about a mathematical situation. Each of the extended constructed-response items was graded using a 5-point rubric specifically designed to fit the demands of the task. A typical problem of this type was the Graphs of Pockets problem discussed in chapter 8.

A summary of student results across all of the extended items revealed that students at grades 4, 8, and 12 were generally unable to make significant headway on such items. Considering such problems at the classroom level necessitates allowing sufficient time for students to think through and communicate their solutions. There was no evidence, however, that the tests were speeded, that is, too short in total time for students to provide complete responses to the tasks. On the contrary, despite detailed instructions, substantial percentages of students appeared at a loss as to how to proceed in answering the extended constructed-response questions on the 1992 NAEP mathematics assessment.

Data collected from teachers at grades 4 and 8 showed that there is a positive relationship between teacher actions and student performance on the extended constructed-response questions, and these results appear in table 2.2. As teachers reported increased emphasis on developing reasoning to solve unique problems, student performance also increased on all forms of assessment items used in NAEP. However, the pattern was not as clear for the effects of teacher emphasis on students' learning to communicate ideas in mathematics effectively. Here the patterns of increased difficulty remained as the problem formats required more student-constructed responses, but the effects on student achievement of the increased

Table 2.2 National Average Percent by Teachers' Reports on the Instructional Emphasis Placed on Reasoning and Communication, Grades 4 and 8

	Developing Reasoning Ability to Solve Unique Problems			Learning How to Communicate Ideas in Mathematics Effectively		
	Heavy Emphasis	Moderate Emphasis	Little or No Emphasis	Heavy Emphasis	Moderate Emphasis	Little or No Emphasis
Grade 4						
Satisfactory or Better on Extended-Response	17	16	14	17	16	15
Correct on Regular Constructed-Response	43	41	40	43	41	42
Correct on Multiple Choice	50	49	48	49	49	50
Grade 8						
Satisfactory or Better on Extended-Response	10	7	2	10	8	6
Correct on Regular Constructed-Response	57	52	44	56	53	53
Correct on Multiple-Choice	60	54	47	59	56	56

Source: Dossey, Mullis, and Jones. *Can Students Do Mathematical Problem Solving? Results from Constructed-Response Questions in NAEP's 1992 Mathematics Assessment,* 1993

emphasis on communication appeared to be negligible at best. Additional analyses of the data presented in table 2.3 showed that twelfth-grade students' success in problem solving increased as students were enrolled in progressively higher-level mathematics coursework. For example, more than 20 percent of students who reported studying calculus scored at the satisfactory level or better on the extended constructed-response question, as opposed to 7 percent of students who reported studying only first-year algebra. Irrespective of the response format, the level of student performance increased as the level of study increased from no mathematics to calculus. This finding reflects the results of the increased opportunity to learn mathematics. This increased opportunity is a result of a number of factors, ranging from increased requirements mandated by state statutes to changes in instruction and the use of technology. Numerous studies that show advanced mathematics coursework to be an important correlate of mathematics achievement strengthen the argument for opportunity to learn as a powerful variable in explaining success in mathematics (Mullis, Jenkins, and Johnson 1994). Nevertheless, year after year, substantial percentages of students disappear from the mathematics pipeline. Only 16 percent of our nation's students graduate from high school having taken precalculus or calculus, and about one-fourth of the eighth-grade students said that if they had a choice they would not take any more mathematics courses (Dossey et al., 1994). Because past experience supports these students' intentions not to take advanced mathematics courses, it is important for parents, teachers, counselors, and others to support and encourage students to continue taking mathematics courses in high school (Lindquist, Dossey, and Mullis 1995).

Possible explanations notwithstanding, the fact remains that, on average, only 16 percent of the fourth-grade students, 8 percent of the eighth-grade students, and 9 percent of the twelfth-grade students provided extended responses to these more thought-provoking problems. The data in table 2.4 show that the performance on these items is considerably lower

Table 2.3 Average Percents of Successful Responses for Different Question Types by Courses Taken in Mathematics, Grade 12

Grade 12	Not Studied	Prealgebra	First-Year Algebra	Second-Year Algebra	Precalculus	Calculus
Satisfactory or Better on Extended-Response	4	3	7	8	13	23
Correct on Regular Constructed-Response	22	26	33	44	54	61
Correct on Multiple-Choice	38	44	50	60	69	74

Source: Dossey, Mullis, and Jones. *Can Students Do Mathematical Problem Solving? Results from Constructed-Response Questions in NAEP's 1992 Mathematics Assessment,* 1993

Table 2.4 Summary of Average Percent Correct by Type of Questions

Grade	Extended Constructed-Response*	Regular Constructed-Response	Multiple-Choice
4	16	42	50
8	8	53	56
12	9	40	56

*Data for the extended constructed-response questions are for the average percents of satisfactory or better responses.

Source: Dossey, Mullis, and Jones. *Can Students Do Mathematical Problem Solving? Results from Constructed-Response Questions in NAEP's 1992 Mathematics Assessment,* 1993

than the performance for each grade group on regular constructed-response items and on multiple-choice items.

As the NAEP assessments move into the future, greater percentages of the tasks presented to students will require constructed responses (NAGB 1996). Table 2.5 presents the composition of recent and projected assessments in terms of the percent of items and the percent of student response time allotted to different item types. These plans reflect intentions to pursue monitoring implementation of some aspects of the NCTM *Standards* in our nation's classrooms.

Table 2.5 Distribution of NAEP Questions and Response Time by Question Type

Question Type	1990		1992		1996	
	Percent of Items	Percent of Time	Percent of Items	Percent of Time	Percent of Items	Percent of Time
Multiple-Choice	75	70	70	65	55	40
Constructed-Response (regular and extended)	25	30	30	35	45	60

The Contexts for Learning Mathematics

Results in NAEP's 1992 reports from students and their teachers suggest no discernible shift to active learning and problem solving. According to students and their teachers, textbooks and worksheets remain the mainstay of instruction (Dossey et al. 1994). Nearly all students solve problems from these sources at least weekly. On the other hand, students report little use of manipulatives (e.g., rulers, geometric shapes, or measuring instruments) and few opportunities to participate in group work.

Considering the emphasis on communication in the NCTM *Standards* documents, teachers and students report a disheartening paucity of attention to written explanations. Only about one-fifth of the students were asked to write even a few sentences about how to solve a mathematics problem as frequently as once a week. Only one-fifth of the students' teachers reported using such techniques as portfolios, projects, and presentations on a monthly basis.

Technology use is creeping into mathematics classrooms but is not widespread. As students progress through school, they increasingly use calculators in mathematics class. Most eighth- and twelfth-grade students (81 and 92 percent, respectively) report having a calculator to use for mathematics, as do nearly one-half of the students in grade 4. Unfortunately, hardly any of the fourth-grade students and fewer than one-third of the eighth-grade students are permitted unrestricted use of calculators in class. Worse yet, opportunities to use calculators appear to be even further curtailed for students in lower-ability classes than for their counterparts in high-ability classes.

Computers are used primarily at grade 4 for playing learning games and for drill and practice. About half of the fourth-grade students use a computer at least some of the time in mathematics class. In contrast, 70 percent or more of the eighth- and twelfth-grade students do so rarely, if ever.

THE NAEP STATE-LEVEL ASSESSMENT

In both 1990 and 1992, state-level replications of the national NAEP assessment in mathematics were carried out in a number of states and territories by using only students attending public schools. These assessments were carried out at grade 8 in 1990 in thirty-seven states, the District of Columbia, Guam, and the Virgin Islands. In 1992, the state-level assessments involved grade 4 and grade 8 and grew to involve forty-one states, the District of Columbia, Guam, and the Virgin Islands. Eighteen of the thirty-seven states and territories that participated in the grade-8 Trial State Assessment Program in both 1990 and 1992 showed significant two-year increases in average mathematics proficiency for their public school students.

As an example of performance at the state level, figure 2.2 shows the differences in average student proficiency existing among the states at grade 8 in 1992. This chart contains 946 simultaneous pairwise comparisons, each adjusted for the fact that multiple comparisons are being made. In the 1992 grade-8 assessment the top-performing states included Iowa, North Dakota, Minnesota, Maine, New Hampshire, Wisconsin, and Nebraska.

Perhaps even more important than illustrating differences in average proficiency, the state-level assessments provided a picture of the considerable variation in performance both across and within states when it comes to student proficiency in mathematics. These differences are driven by a large number of social, educational, and resource factors that control students' access to opportunities for maximizing their learning in mathematics. The increases in mathematics proficiency between 1990 and 1992 for the nation and in many states did little to alter the relative standings of the various demographic groups. The results for demographic variables within states tended to show about the same patterns for student achievement seen at the national level.

There has been an interest in linking state-level NAEP results to results from international mathematics assessments. In 1990–91, as part of the

Instructions: Read down the column directly under a state name listed in the heading at the top of the chart. Match the shading intensity surrounding a state postal abbreviation to the key below to determine whether the average mathematics performance of this state is higher than, the same as, or lower than the state in the column heading.

State has statistically significantly higher average proficiency than the state listed at the top of the chart

No statistically significant difference from the state listed at the top of the chart

State has statistically significantly lower average proficiency than the state listed at the top of the chart

The between state comparisons take into account sampling and measurement error and that each state is being compared with every other state. Significance is determined by an application of the Bonferroni procedure based on 946 comparisons by comparing the difference between the two means with four times the square root of the sum of the squared standard errors.

Fig. 2.2 Comparisons of overall mathematics average proficiency 1992, grade 8

Source: Mullis et al. *NAEP 1992 Mathematics Report Card for the Nation and States,* 1993, p. 79

second International Assessment of Educational Progress (IAEP), a total of twenty countries surveyed the mathematics performance of 13-year-old students (Lapointe, Mead, and Askew 1992). There have been studies conducted to link the IAEP data to the data obtained from the NAEP state-level assessments (Beaton and Gonzalez 1993; Pashley and Phillips 1993). These studies suggest that some students in the highest achieving states on the NAEP assessment perform as well as the best students in some foreign countries having high performance levels in school mathematics. These studies have provided the groundwork for plans to link the data obtained from the Third International Mathematics and Science Study (TIMSS) with the data from the 1996 state-level NAEP assessments. With forty-five countries participating, five grades assessed in two school subjects, more than half a million students tested in more than thirty languages, and millions of open-ended responses generated, TIMSS is perhaps the most ambitious study of comparative education ever conducted. Publication of the TIMSS results will begin in late 1996 with a comparative curriculum study and achievement results for grades 7 and 8.

THE NAEP TREND ASSESSMENT

Increases and decreases in student performance in academic subjects have become as important in many individuals' minds as rises and dips in the Dow Jones Index, and a third NAEP feature includes the long-term trend assessments carried out in 1973, 1978, 1982, 1986, 1990, and 1992. Across the years, the NAEP trends in students' mathematics proficiency have been of particular interest to policymakers, the news media, and others because mathematics serves as the key to related performance in many of the sciences and other areas in the curriculum.

Unlike the recent national NAEP assessments, the NAEP trend assessments continue NAEP's initial data-gathering process of sampling students at ages 9, 13, and 17. Another difference is that the trend assessments employ a different set of items, reflecting a more conservative view of the school curriculum—one more reflective of the 1970s curriculum than of contemporary goals in mathematics education, such as the problem-solving analyses associated with the 1992 national and state assessments. Nevertheless, the importance of student proficiency with mathematics fundamentals has not diminished. The expectations are for students to increase their proficiency in basic skills and also reach expectations associated with problem solving and application.

The line graphs in figure 2.3 depict the overall trends in mathematics achievement between 1973 and 1992 (Mullis et al. 1994). Average mathematics proficiency improved between 1973 and 1992 at ages 9 and 13. The data for age 17 show declines in performance between 1973 and 1982 fol-

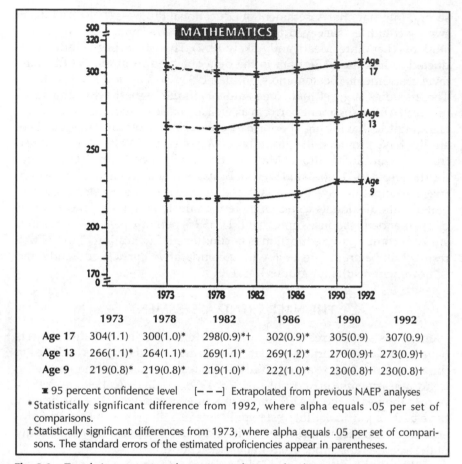

	1973	1978	1982	1986	1990	1992
Age 17	304(1.1)	300(1.0)*	298(0.9)*†	302(0.9)*	305(0.9)	307(0.9)
Age 13	266(1.1)*	264(1.1)*	269(1.1)*	269(1.2)*	270(0.9)†	273(0.9)†
Age 9	219(0.8)*	219(0.8)*	219(1.0)*	222(1.0)*	230(0.8)†	230(0.8)†

✖ 95 percent confidence level   [– – –] Extrapolated from previous NAEP analyses
*Statistically significant difference from 1992, where alpha equals .05 per set of comparisons.
†Statistically significant differences from 1973, where alpha equals .05 per set of comparisons. The standard errors of the estimated proficiencies appear in parentheses.

Fig. 2.3 Trends in average mathematics proficiency for the nation, 1973 to 1992
Source: Mullis et al. *NAEP 1992 Trends in Academic Progress,* 1994, p. 76

lowed by recovery. By 1992, performance had returned to the initial 1973 level. The gains during the decade between 1982 and 1992 were notable at all three ages. Although the two-year trends between 1990 and 1992—stability at age 9 and slight gains at ages 13 and 17—do not show dramatic gains, they do indicate continued progress in recent assessments. To summarize, performance on mathematics fundamentals was at least as high at age 17 as it was in the early 1970s, and higher at ages 9 and 13.

Statistical analyses of the nature of the trend data across the full set of assessments indicate significant positive linear changes in student proficiency for ages 9, 13, and 17. These findings support the claim that long-term improvement is indeed taking place in our students' performance in mathematics, at least as measured by the NAEP trend assessments.

Students in the racial/ethnic groups showed improvement at each of the three age levels, with the exception of White 17-year-olds. Despite progress in reducing the performance differences across the past two decades, however, the gaps remain wide. In 1992, both Black and Hispanic students, on average, demonstrated significantly lower proficiency than White students. Both males and females showed improvement at ages 9 and 13. Between 1973 and 1992, a slight advantage favoring females at age 9 reversed to a slight advantage favoring males. A similar pattern was observed at age 13. At age 17, there appeared to be a slight narrowing of the gender gap.

Information about student performance at various levels on the NAEP trend scales is available back to 1978. Greater percentages of 9-, 13-, and 17-year-olds demonstrated an understanding of the mathematics assessed by NAEP in 1992 than in 1978.

CONCLUSION

The trend data reveal that in the long term, our students continue to maintain or improve their mastery of basic objectives for mathematics education. Combined with data from the 1990 and 1992 national and state-level assessments, which place more emphasis on problem solving and calculator use, the NAEP data suggest that improvement is really under way in American mathematics education. Students are not forgetting the traditionally held concepts and skills measured in greater proportion by the NAEP trend examinations while they focus their studies more heavily on a broader view of mathematics and employ technology in doing so. At the same time, we note that there is still distance to cover in assisting students to move from their present levels of performance to higher levels on both basic objectives and those involving problem solving and the use of technology. If we as a nation are to achieve the goals set in the NCTM *Standards* or by state and local school authorities, educators, parents, and policymakers will have to continue their emphasis on mathematics and its importance, support and encourage students in their efforts, and maintain focus on the objectives for improvement.

REFERENCES

Beaton, Albert E., and Eugenio J. Gonzalez. "Comparing the NAEP Trial State Assessment Results with the IAEP International Results." In *Setting Performance Standards for Student Achievement: Background Studies*, pp. 371–98. Stanford, Calif.: National Academy of Education, 1993.

Dossey, John A., Ina V. S. Mullis, Steven Gorman, and Andrew S. Latham. *How School Mathematics Functions: Perspectives from the NAEP 1990 and 1992 Assessments*. Washington, D.C.: National Center for Education Statistics, October 1994.

Dossey, John A., Ina V. S. Mullis, and Chancey O. Jones. *Can Students Do Mathemati-cal Problem Solving? Results from Constructed-Response Questions in NAEP's 1992 Mathematics Assessment.* Washington, D.C.: National Center for Education Sta-tistics, 1993.

Lapointe, Archie E., Nancy A. Mead, and Janice M. Askew. *Learning Mathematics.* Princeton, N.J.: International Assessment of Educational Progress, Educational Testing Service, 1992.

Lindquist, Mary M., John A. Dossey, and Ina V. S. Mullis. *Reaching Standards: A Progress Report on Mathematics.* Princeton, N.J.: Policy Information Center, Edu-cational Testing Service, 1995.

Mullis, Ina V. S. *The NAEP Guide: A Description of the Content and Methods of the 1990 and 1992 Assessments.* Washington, D.C.: National Center for Education Statis-tics, 1991.

Mullis, Ina V. S., John A. Dossey, Jay R. Campbell, Claudia A. Gentile, Christine O'Sullivan, and Andrew S. Latham. *NAEP 1992 Trends in Academic Progress.* Washington, D.C.: National Center for Education Statistics, 1994.

Mullis, Ina V. S., John A. Dossey, Eugene H. Owen, and Gary W. Phillips. *NAEP 1992 Mathematics Report Card for the Nation and the States: Data from the National and Trial State Assessments.* Washington, D.C.: National Center for Education Sta-tistics, 1993.

Mullis, Ina V. S., Frank Jenkins, and Eugene G. Johnson. *Effective Schools in Math-ematics: Perspectives from the NAEP 1992 Assessment.* Washington, D.C.: National Center for Education Statistics, 1994.

National Assessment of Educational Progress. *Mathematics Objectives: 1990 Assess-ment.* Princeton, N.J.: Educational Testing Service, National Assessment of Edu-cational Progress, 1988.

National Assessment Governing Board. *Setting Appropriate Achievement Levels for the National Assessment of Educational Progress: Policy Framework and Technical Pro-cedures.* Washington, D.C.: National Assessment Governing Board, 1990.

_____. *Mathematics Framework for the 1996 National Assessment of Educational Progress.* Washington, D.C.: National Assessment Governing Board, 1996.

National Council of Teachers of Mathematics. *Curriculum and Evaluation Standards for School Mathematics.* Reston, Va.: National Council of Teachers of Mathemat-ics, 1989.

_____. *Professional Standards for Teaching Mathematics.* Reston, Va.: National Council of Teachers of Mathematics, 1991.

_____. *Assessment Standards for School Mathematics.* Reston, Va.: National Council of Teachers of Mathematics, 1995.

Pashley, Peter J., and Gary W. Phillips. *Toward World-Class Standards: A Research Study Linking International and National Assessments.* Princeton, N.J.: Educational Testing Service, 1993.

Phillips, Gary W., Ina V. S. Mullis, Mary Lyn Bourque, Paul L. Williams, Ronald K. Hambleton, Eugene H. Owen, and Paul E. Barton. *Interpreting NAEP Scales.* Washington, D.C.: National Center for Education Statistics, 1993.

3

NAEP Findings regarding Race/Ethnicity and Gender: Affective Issues, Mathematics Performance, and Instructional Context

Edward A. Silver, Marilyn E. Strutchens, & Judith S. Zawojewski

THE PERFORMANCE of students on the NAEP mathematics assessment is reported not only for the nation as a whole but also for subgroups stratified by various demographic factors, including gender, race/ethnicity, and community type (in part a socioeconomic indicator). Thus, it is possible for example to use results from the 1992 NAEP mathematics assessment to examine and compare the performance of males and females or that of students who are members of various race/ethnicity subgroups. In fact, since its inception, NAEP has been a primary source of information about the relative performance of these groups of students.

As we shall see in this chapter, there is a clear pattern of differential mathematics performance among White, Black, and Hispanic students, but there is little difference across groups in students' attitudes toward mathematics. Comparisons can be readily made regarding the relative performance of students using race/ethnicity categories, but interpretations of any observed similarities or differences must be made with care because of the complex factors that are glossed over in such comparisons. Race/

At the direction of the U.S. Office of Education, NAEP uses five major race/ethnicity demographic categories: White, Black, Hispanic, Asian/Pacific Islander, and American Indian. For some NAEP analyses the numbers of Asian/Pacific Islander and American Indian students is too small to allow sufficient confidence in the generalizability of the findings. Therefore, in this chapter we omit these two groups from our discussion and focus only on the first three categories. Also, although we recognize that there is a diversity of opinion about the appropriate terms to be used when referring to members of the diverse U.S. population racial and ethnic subgroups, here we use the designations Black, White, and Hispanic because they are used by NAEP. NAEP's style of capitalization has been followed for terms designating ethnic and racial groups.

ethnicity category designations mask a variety of ethnic, school, and societal factors that may differentially affect the mathematics learning of stu-

HIGHLIGHTS

- At all grade levels, there was little difference in the overall performance of males and females, and there was no gender difference in performance within the three major race/ethnicity subgroups—White, Black, and Hispanic. Likewise, there was little overall difference between males and females in the reported frequency of taking core college preparatory courses, with the exception of calculus, which was taken more frequently by males.

- Between 1973 and 1992 all students gained in proficiency on test items that assess fairly traditional mathematical concepts and skills, and Black and Hispanic students narrowed the performance gap with White students on these items. Nevertheless, the 1992 NAEP results indicate that substantial performance gaps still remain and that gaps in relative performance are especially evident for more complex items, such as extended constructed-response tasks.

- At all grade levels, the percent of students judged to have attained a Proficient or Advanced level of performance was very small, and it was alarmingly low for Black and Hispanic students. White students performed significantly better than Black or Hispanic students, and Hispanic students generally performed better than Black students. Differences between groups stratified by socioeconomic factors were larger than those between groups stratified by race/ethnicity.

- Black and Hispanic students at grade 8 are more likely than White students to declare an intention not to study mathematics in the future and are less likely to study algebra or prealgebra. In high school, Black and Hispanic students study mathematics for less time, are more likely to take general mathematics or consumer mathematics, and are less likely to take advanced college preparatory mathematics courses like algebra 2 and precalculus.

- At grades 4 and 8, there were relatively few major differences in the reported teaching practices for teachers of White, Black, and Hispanic students. One notable exception was the tendency of teachers of fourth- and eighth-grade Black and Hispanic students to report using multiple-choice assessments far more frequently than teachers of White students.

- Regardless of gender or race/ethnicity, students expressed generally similar and positive attitudes toward mathematics and views of mathematics learning and its social or economic utility. But a substantial percent of students at all grade levels view the learning of mathematics as primarily involving memorization.

dents within and across these groups. Ethnic factors include one's language, religion, and distinctive customs (Banks 1989). School and societal factors include academic "tracking" policies that affect access to mathematics courses, teachers' beliefs about students, assessment or instructional practices and the availability of appropriate resources, parents' involvement and expectations, students' self-perceptions and expectations, and socioeconomic status (Oakes 1990a).

Although NAEP is not able to provide a full account of the ethnic, school, and social context for these performance differences, the data do suggest that factors related to opportunity to learn (e.g., patterns of mathematics course taking, classroom instruction and assessment, and socioeconomic status) may be important contributors to the observed differential performance of students across the race/ethnicity groups. These factors are themselves interrelated. In particular, because contemporary urban schools reflect a legacy derived from a persistent historical association between race and poverty, there is a confounding of these factors in the NAEP data. A disproportionate number of students in poor urban communities are Black or Hispanic. For example, in the NAEP Disadvantaged Urban sample in 1992 about 68 percent of the students in grade 4 were Black or Hispanic, and only 28 percent were White. In contrast, 84 percent of the students in the Advantaged Urban sample were White, and only 12 percent were Black or Hispanic. Therefore, although we use the race/ethnicity categories as the primary reporting mode in this chapter, the reader is advised to consider the broad range of associated factors, with special attention to poverty, before drawing conclusions from these findings.

STUDENTS' ATTITUDES AND BELIEFS

Students' mathematics achievement can be influenced by affective factors, such as students' beliefs, attitudes, and emotions about mathematics and about themselves as mathematics learners. For this reason, many researchers have examined relationships between mathematical performance and various affective factors, such as students' confidence in their ability to learn mathematics, their beliefs about mathematics, and their perceptions of the utility of mathematics (Johnson 1984; Matthews 1984; Reyes and Stanic 1988). Affective factors have also been shown to be important determinants of students' decisions to enroll in mathematics courses. In particular, Reyes (1980) reported that students' decisions concerning enrollment and effort in mathematics courses were influenced by their attitudes and beliefs. Beliefs and attitudes also appear to influence achievement in more subtle ways. For example, some research suggests that students' attitudes toward mathematics influence the amount of attention they receive from their teachers. Reyes (1980) also found that teachers appear to pay more attention to students who exhibit high mathematical confidence than they do to students who are less confident, even when both sets of students perform equally well in mathematics.

Because of the potential impact that affective factors can have on student achievement, the mathematics attitudes of students from different gender and race/ethnicity groups have been examined in the past, and NAEP has served as one important source of this information (Hart 1989; Johnson 1989; Matthews et al. 1984; Meyer 1989; Stanic and Hart 1995). In the 1992 NAEP mathematics assessment, as in previous NAEP assessments, students were given questionnaires to determine their attitudes and beliefs related to mathematics. Seven statements were given to students at all three grade levels, and they were asked to indicate the extent to which they agreed with each statement. Students in grades 8 and 12 selected one of five choices (Strongly Agree, Agree, Undecided, Disagree, or Strongly Disagree); students in grade 4 selected one of three choices (Agree, Undecided, or Disagree).

For three statements regarding students' attitudes toward mathematics and their perceptions of themselves as learners of mathematics, table 3.1 gives a summary of the percent of students who agree (for students in grades 8 and 12 this reflects a sum of the percent selecting Agree and the percent selecting Strongly Agree). As has been found in prior NAEP assessments, older students were less likely than younger students to agree that they like mathematics and to express confidence in their ability to do it, and there was little difference among White, Black, and Hispanic students in their responses.

On the 1992 NAEP assessment, more than two-thirds of the fourth-grade students agreed that they liked mathematics and a similar proportion agreed that they were good at it, but only about one-half of the twelfth-grade students expressed these views. Males in grades 8 and 12 were significantly more likely than females to agree that they liked mathematics, but there was little or no difference between males and females in their perceptions of confidence in being good at mathematics. The other attitude/belief item suggests one plausible explanation for students' decreasing affection for mathematics and their diminished sense of confidence over the years. Older students were less likely than younger students to indicate that they understood classroom mathematics instruction. This pattern was observed across all race/ethnicity and gender categories.

Three other statements assessed students' attitudes and beliefs about the nature of mathematics learning and the utility of school mathematics. A summary of the percent of students who agreed with each of these statements is given in table 3.2.

As has been noted in prior NAEP assessments, a large percentage of students appear to view the learning of mathematics as primarily involv-

Table 3.1   Percent of Students Agreeing with Statements regarding Perceptions of Themselves with Respect to Mathematics

Statement	Grade	Percent Agreeing					
		Nation	White	Black	Hispanic	Male	Female
I like mathematics.	4	71	71	74	72	72	71
	8	60	61	62	51	66	54
	12	51	51	50	46	56	45
I am good at mathematics.	4	65	66	64	61	70	68
	8	57	57	64	55	60	55
	12	51	49	55	55	53	48
I understand most of what goes on in math class.	4	80	82	75	73	80	79
	8	80	81	81	73	83	78
	12	66	66	71	64	69	64

Source: National Center for Education Statistics. *Data Compendium for the NAEP 1992 Mathematics Assessment of the Nation and the States,* 1993

ing memorization, with younger students more likely than older students to hold this view. An instructional emphasis on facts and procedures in the elementary grades appears to dominate the perceptions of a majority of students in grade 4, whereas students in grade 8 and especially grade 12 are more likely to have encountered mathematics that calls for more-complex thinking. Nevertheless, it is troubling to note nearly two-thirds of Black students, even in grades 8 and 12, view the learning of mathematics as primarily involving memorization, which suggests that their experience with instruction emphasizing rote learning rather than problem solving or critical thinking is more likely to extend to grades 8 and 12 than is the experience of White students. Also disturbing is the fact that the proportion of Hispanic students agreeing with this view is higher than that of White students in grades 8 and 12, which probably also reflects the low-level cognitive demands of the instruction to which they have been exposed.

Despite the finding that many students associate mathematics with memorization, students at all grade levels appear to view mathematics as

Table 3.2 Percent of Students Agreeing with Statements regarding the Nature and Utility of Mathematics

Statement	Grade	Percent Agreeing					
		Nation	White	Black	Hispanic	Male	Female
Learning mathematics is mostly memorizing facts.	4	57	55	64	51	57	56
	8	44	38	66	57	45	42
	12	41	35	66	44	42	39
Mathematics is useful for solving everyday problems.	4	66	67	63	61	67	66
	8	81	81	82	80	82	80
	12	71	68	76	73	72	69
Almost all people use math in their jobs.	4	74	76	70	67	74	74
	8	87	87	90	87	87	88
	12	74	70	81	77	76	73

Source: National Center for Education Statistics. *Data Compendium for the NAEP 1992 Mathematics Assessment of the Nation and the States,* 1993

having considerable social and economic utility. Across the grades assessed in NAEP a substantial majority of students agreed that mathematics is useful for solving everyday problems and that it is used by workers in their jobs. At all grade levels, and for all race/ethnicity and gender subgroups, students were somewhat more likely to view mathematics as more useful in work than in the rest of life. These findings are similar to those reported by Johnson (1989) in his review of data from the 1986 NAEP mathematics assessment.

Another attitude/belief statement that students at all grades were given was, Mathematics is more for boys than for girls. Overall, more than three-fourths of students at all grade levels disagreed (that is, disagreed at grade 4 and disagreed or strongly disagreed at grades 8 and 12), and the strongest disagreement was found in grade 8. At grades 8 and 12, the rate of disagreement was similar for White, Black, and Hispanic students; at grade 4, White students were more likely to disagree than either Black or Hispanic students. In contrast to the general similarity across race/ethnicity groups, there was a significant difference in how males and females responded to this statement at all grade levels. At grades 4, 8, and 12, the percent of females who disagreed was 84, 91, and 88, respectively; whereas the percents were only 64, 78, and 67, respectively, for males. Thus, although the vast majority of female students may not see mathematics as a male domain, considerably more of their male counterparts do view it in this way.

STUDENT PERFORMANCE AND TRENDS OVER TIME

Although affective factors are thought by many to be related to student performance, the relationship is neither simple nor direct. In fact, in the NAEP data we note that the observed patterns of difference and nondifference for the affective items are different from the patterns observed in the actual performance of students on the NAEP mathematics assessment for the gender and race/ethnicity subgroups. Although there was essentially no difference across the race/ethnicity groups on attitude items, the performance differences are substantial. For males and females the reverse trend was found: some attitude differences but almost no performance differences.

Mathematical Proficiency in 1992

As explained in the chapter by Dossey and Mullis, student performance on the 1992 NAEP Mathematics Assessment was expressed in several ways, including both a proficiency scale score (with scores ranging from 0 to 500) and three achievement levels (*Basic, Proficient, Advanced*). Table 3.3 provides the average proficiency score and the percent of students classified at the Proficient or Advanced achievement levels (i.e., at or above the Proficient achievement level) for all males and females in the sample and also for males and females within each race/ethnicity subgroup. At all grade levels the proportion of all students classified as being at or above the Proficient achievement level is quite small—only about one in five students at grades 4 and 12 and about one in three at grade 8. Moreover, the proportion of Black and Hispanic students so classified is alarmingly low; no more than one of every ten Black or Hispanic students at any grade level is classified as being at or above the Proficient achievement level, and the proportions are often as low as one in twenty.

Two further observations can be made about the information reported in table 3.3. First, there is little difference between gender subgroups at any grade level. The performance of males and females within the total NAEP sample was quite similar at each grade level, as was the performance of males and females within each of the race/ethnicity categories (i.e., Black males and females performed equally well, as did Hispanic males and

WILLIAM WOODS UNIVERSITY LIBRARY

Table 3.3 Average Proficiency and Percent of Students Classified at or above Proficient Achievement Level

			Average Proficiency	Percent at or above Proficient
Grade 4	**Nation**	Male	220(0.8)	20(1.1)
		Female	217(1.0)	17(1.3)
	White	Male	228(1.0)	25(1.5)
		Female	225(1.4)	21(1.8)
	Black	Male	192(1.6)	4(1.0)
		Female	191(1.6)	2(0.8)
	Hispanic	Male	200(1.6)	5(1.6)
		Female	201(1.7)	6(1.3)
Grade 8	**Nation**	Male	267(1.1)	25(1.3)
		Female	268(1.0)	24(1.3)
	White	Male	277(1.2)	32(1.5)
		Female	277(1.1)	31(1.6)
	Black	Male	237(1.9)	4(1.3)
		Female	237(1.5)	3(0.8)
	Hispanic	Male	246(1.1)	9(1.4)
		Female	247(1.0)	8(1.3)
Grade 12	**Nation**	Male	301(1.1)	25(1.3)
		Female	297(1.0)	24(1.3)
	White	Male	307(1.0)	25(1.3)
		Female	303(1.0)	24(1.3)
	Black	Male	277(2.3)	9(1.4)
		Female	273(1.8)	8(1.3)
	Hispanic	Male	281(3.4)	9(1.4)
		Female	285(2.5)	8(1.3)

Source: National Center for Education Statistics. *Data Compendium for the NAEP 1992 Mathematics Assessment of the Nation and the States,* 1993

WILLIAM WOODS UNIVERSITY LIBRARY

females and also White males and females). Similarly, the percent of males and females classified at or above the Proficient achievement level is similar for all groups. Second, significant differences exist among race/ethnicity subgroups at each grade level. The average proficiency of White students was considerably higher than that of Black or Hispanic students at all grade levels, and Hispanic students performed better than Black students at grades 4 and 8. There is a similar pattern of variation in the percent of White, Black, and Hispanic students classified at or above the Proficient achievement level at each grade level.

The general patterns of performance noted for the overall results in table 3.3 were mirrored to a great extent in performance within each mathematical content area (numbers and operations, geometry, and so on). That is, in each content area and at each grade level, White students performed significantly better than Black and Hispanic students, and Hispanic students performed better than Black students. These NAEP findings suggest that a substantial gap exists between the performance of White students and that of Black and Hispanic students and that this performance gap persists across all areas of the mathematics curriculum. A significant difference is also found between Hispanic and Black students at all grade levels; however, caution is advised in generalizing the results from NAEP's tested sample of Hispanic students to the entire population. At least some Hispanic students were among the approximately 3 percent of the school population that was eligible to be tested in the 1992 NAEP mathematics assessment but excluded because the students were judged by their teachers or by school or district administrators to lack sufficient proficiency with the English language (Spencer 1994).

With respect to gender, little difference in performance was found for males and females on the overall assessment; similarly, the examination of performance in specific content areas found little difference in the performance of males and females overall or within race/ethnicity subgroups. The only significant overall gender difference appeared at grade 12 for items in the areas of measurement and geometry, with males having significantly better than average performance in those content areas. However, it is interesting to note that almost all of this difference was due to differential performance by White males and females; no significant difference was found for these content areas in the performance of Black or Hispanic males and females.

Students' Mathematical Proficiency, 1990–1992

Table 3.4 contains average proficiency scores obtained on the 1990 and 1992 NAEP mathematics assessments at each of the three grade levels reported for various demographic subgroups on a common set of items administered in both 1990 and 1992. Looking at the performance of the three race/ethnicity subgroups, in both years we see that White students obtained significantly higher proficiency scores at all grade levels than Black and Hispanic students. Moreover, White students obtained significantly higher proficiency scores at all grade levels in 1992 than in 1990, whereas in 1992 Black and Hispanic students performed significantly better only in grade 12.

Table 3.4 Average Mathematics Proficiency by Race/Ethnicity, Community Type, and Gender

		Grade 4	Grade 8	Grade 12
White	1992	227(0.9)>	277(1.0)>	305(0.9)>
	1990	220(1.1)	270(1.4)	300(1.2)
Black	1992	192(1.3)	237(1.4)	275(1.7)>
	1990	189(1.8)	238(2.7)	268(1.9)
Hispanic	1992	201(1.4)	246(1.2)	283(1.8)>
	1990	198(2.0)	244(2.8)	276(2.8)
Advantaged Urban	1992	237(2.1)	288(3.6)	316(2.6)
	1990	231(3.0)	280(3.2)	306(6.2)
Disadvantaged Urban	1992	193(2.8)	238(2.6)<	279(2.4)
	1990	195(3.0)	249(3.8)	276(6.0)
Male	1992	220(0.8)>	267(1.1)>	301(1.1)>
	1990	214(1.2)	263(1.6)	297(1.4)
Female	1992	217(1.0)>	268(1.0)>	297(1.0)>
	1990	212(1.1)	262(1.3)	292(1.3)

> The value for 1992 is significantly higher than the value for 1990 at about the 95 percent confidence level.

< The value for 1992 is significantly lower than the value for 1990 at about the 95 percent confidence level.

Note: The standard errors of the estimated proficiencies appear in parentheses.

Source: Mullis et al. *NAEP 1992 Trends in Academic Progress,* 1994

Although the pattern of differences among the race/ethnicity group students is clear, an examination of the data in table 3.4 also reveals that the size of performance gap between White and either Black or Hispanic students was smaller than the performance gap between students in NAEP's Disadvantaged Urban and Advantaged Urban samples at all grade levels. For example, for eighth-grade students in 1992, the difference between White and Black students' proficiency was 40 points (277 – 237), and the difference between the Advantaged and Disadvantaged Urban groups was 50 points (288 – 238). The size of the gap in proficiency scores associated with differences in race/ethnicity at any grade level is smaller than the size of the corresponding proficiency score gap associated with differences in community type, which is a socioeconomic indicator. Given the apparent importance of socioeconomic factors in affecting student performance, it is disturbing that the proficiency of eighth-grade Disadvantaged Urban students significantly declined in 1992—the only significant decline noted in table 3.4.

As would be expected from data reported previously in table 3.3, the performance of males and females in 1992 was quite similar at all grade levels, with males performing slightly better at grades 4 and 12 and females at grade 8 but with very small performance differences overall. It is also encouraging to note that the performance of males and females increased significantly between 1990 and 1992 at all grade levels, with females making slightly greater gains at grades 8 and 12.

Long-Term Trends in Students' Mathematical Proficiency, 1973–1992

In addition to looking at short-term trends such as those from 1990 to 1992, NAEP has also examined long-term trends in student performance by administering the same (or very similar) sets of items to age-based samples of students since 1978. Performance for NAEP in 1973 was obtained by extrapolating from performance on subsequent assessments. Because the set of items on which this analysis is based has been kept secure and has remained relatively stable over time (and constant since 1986), it contains items that can generally be characterized as assessing fairly basic and traditional mathematics concepts and skills (Mullis et al. 1994). Table 3.5 provides the average proficiency score for White, Black, and Hispanic students at each of three age levels for all six NAEP mathematics administrations.

On the basis of the data presented in table 3.5, two observations can be made. First, 9-year-old and 13-year-old students in these three race/ethnicity subgroups performed better on average in 1992 than they did in 1973; and 17-year-old Black and Hispanic students also performed significantly

Table 3.5 Average Mathematical Proficiency of White, Black, and Hispanic Students, 1973–1992

		White	Black	Hispanic
Age 9	1973	225(1.0)*	190(1.8)*	202(2.4)*
	1978	224(0.9)*	192(1.1)*	203(2.3)*
	1982	224(1.1)*	195(1.6)*	204(1.3)*
	1986	227(1.1)*	202(1.6)†	205(2.1)
	1990	235(0.8)†	208(2.2)†	214(2.1)†
	1992	235(0.8)†	208(2.0)†	212(2.3)†
Age 13	1973	274 (0.9)*	228(1.9)*	239(2.2)*
	1978	272(0.9)*	230(1.9)*	238(2.2)*
	1982	274(1.0)*	240(1.6)*†	252(1.6)*†
	1986	274(1.3)*	249(2.3)†	254(2.9)†
	1990	276(1.1)	249(2.3)†	255(1.8)†
	1992	279(0.9)†	250(1.9)†	259(1.8)†
Age 17	1973	310(1.0)	270(1.3)*	277(2.2)*
	1978	306(0.9)*†	268(1.3)*	276(2.2)*
	1982	304(0.9)*†	272(1.3)*	277(2.0)*
	1986	308(1.0)*	279(2.1)†	283(2.9)
	1990	310(1.0)	288(2.8)†	284(2.9)
	1992	312(0.8)	286(2.2)	292(2.6)†

* significant difference w/1992
† significant difference w/1973
Note: The standard errors of the estimated proficiencies appear in parentheses.
Source: Mullis et al. *NAEP 1992 Trends in Academic Progress,* 1994

better in 1992 than in 1973. (See chapter 2 for a different presentation and discussion of these data.) Contrary to the oft-stated claim that students are less mathematically skilled now than in the past, these data suggest that the performance of all students on the traditional mathematical concepts and skills assessed in this portion of the NAEP mathematics assessment was actually significantly better in 1992 than in 1973, with the lone exception being the performance of 17-year-old White students. Second, Black and Hispanic students made substantially greater performance gains than White students during the period 1973–1992. Thus, Black and Hispanic students have considerably narrowed the performance gap with White students at all age levels. Nevertheless, substantial performance gaps remain on these tasks assessing traditional topics, and differences in performance are even more evident when newer types of tasks are considered, such as the extended constructed-response tasks used on NAEP for the first time in 1992.

Student Performance on Three Types of NAEP Items in 1992

Table 3.6 summarizes the performance of students in 1992 on the three types of tasks included in the NAEP mathematics assessment at all three grade levels for various demographic subgroups. As would be expected from data reported in previous sections of this chapter, White students performed considerably better than Black or Hispanic students at each grade level on all three item types, and there is little or no difference between males and females on any item type at any grade level, except for a slight advantage for females on extended constructed-response tasks at grade 8. A closer examination of the data in table 3.6, however, sheds light on the need for caution in interpreting the apparently positive news from the long-term trend data reported earlier in table 3.5, indicating a narrowing gap in performance differences among White, Black, and Hispanic students.

Because students performed less well on extended constructed-response tasks than on regular constructed-response items (which in turn had somewhat lower rates of success than multiple-choice items), the absolute differences in performance between White and Black or Hispanic students are smaller for the extended tasks than for the multiple-choice items. Nevertheless, a consideration of the *relative* performance of students in the three groups suggests that the difficulties are more pronounced on the more complex, extended tasks than on the simpler, multiple-choice tasks. Consider, for example, the relative performance of White and Black students on the three item types. If we ignore the standard errors, which are about the same size for the two subgroups, the ratios of Black to White student performance for multiple-choice tasks, regular constructed-response items, and extended constructed-response tasks in grade 8 are 0.70, 0.61, and 0.20,

Table 3.6 Correct Response Rate for Three NAEP Item Formats by Race/Ethnicity, Community Type, and Gender Categories

	Multiple-Choice Average Percent Correct			Regular Constructed-Response Average Percent Correct			Extended Constructed-Response Average Satisfactory or Better		
	4	8	12	4	8	12	4	8	12
Nation	50	56	56	42	53	40	16	8	9
White	53	60	59	47	59	44	20	10	10
Black	38	42	46	24	36	26	5	2	4
Hispanic	42	46	49	31	42	32	7	3	4
Advantaged Urban	59	65	65	54	64	49	26	16	13
Disadvantaged Urban	38	43	48	26	37	30	5	3	5
Male	50	56	58	43	53	41	16	7	8
Female	48	56	55	41	54	40	17	10	9

Source: Dossey, Mullis, and Jones. *Can Students Do Mathematical Problem Solving? Results from Constructed-Response Questions in NAEP's 1992 Mathematics Assessment,* 1993

respectively. That is, in 1992 Black eighth-grade students performed 70 percent as well as White eighth-grade students on multiple-choice items, but they performed only 20 percent as well on extended constructed-response tasks. The performance ratios for other grade levels or for White and Hispanic students indicate a similar trend. In all cases, the relative performances are much more alike for the multiple-choice and regular constructed-response items than for the extended constructed-response tasks.

Although the long-term trend data reported in table 3.5 suggested that the performance gap has been closing over time between White and Black or Hispanic students on items that assess basic-level knowledge and skills, the data from the 1992 NAEP assessment reported in table 3.6 suggest the existence of potentially perilous performance differences on more complex, extended tasks. In fact, these differences could lead to a widening of the performance gap in the future as complex, extended tasks become more prevalent in the NAEP assessment. Moreover, a similar pattern (but with smaller performance ratios) is found when one compares the performance of students in the Disadvantaged Urban and Advantaged Urban samples. For example, the ratios of Disadvantaged Urban to Advantaged Urban student performance for multiple-choice tasks, regular constructed-response items, and extended constructed-response tasks in grade 8 are 0.66, 0.58, and 0.19, respectively. In the next section we examine a variety of other data from the 1992 NAEP mathematics assessment in order to discern the instructional context associated with these performance results and observed differences.

INSTRUCTIONAL CONTEXT

Students' performance on the NAEP mathematics assessment is almost certainly influenced by the opportunities they have had in their study of mathematics. Several researchers (e.g., Epstein and MacIver 1992; Welch, Anderson, and Harris 1982) have tried to establish a direct relationship between course taking and proficiency, and the general finding in these efforts has been the existence of a positive association. Perhaps even more compelling is the work of others who have examined a related proposition, namely, that differences in mathematical proficiency between groups of students are due to the inequitable distribution of opportunities to learn mathematics (e.g., Oakes 1990b). Given the existence in NAEP of substantial performance differences among White, Black, and Hispanic students and given the general nonexistence of differences between males and females, it may be useful to consider how data regarding instructional context relate to the findings regarding mathematical proficiency.

Data regarding the instructional context in which students study mathematics in U.S. schools are collected through questionnaires administered to students, teachers, and school administrators as part of the NAEP assessment. These data illuminate not only the general instructional conditions associated with mathematics education in the nation (see chapter 4) but also the particular instructional conditions that undoubtedly influence the observed performance trends for students in the race/ethnicity and gender categories.

Students' Mathematics Course Enrollment

One likely source of differential mathematical proficiency is differences in the amount of mathematics studied (Welch, Anderson, and Harris 1982). Enrollment in any mathematics course presumably affords students an opportunity to learn some mathematics. Data collected as part of the NAEP mathematics assessment afford a view of mathematics course-taking patterns by the nation's students. In fact, several analyses have been conducted using NAEP data to relate overall findings regarding students' course taking to findings regarding mathematical proficiency. For example, Dossey et al. (1994) found a clear, direct relationship between the amount of mathematics studied in high school and the level of performance on the NAEP mathematics assessment for students in grade 12. A subsequent analysis by Naifeh and Shakrani (1996) indicated that the mathematical proficiency of twelfth-grade students is linked not only to the amount of mathematics studied but also to what mathematics is studied (that is, to the specific courses that students take in high school). In particular, Naifeh and Shakrani found that student proficiency was directly related to the number of mathematics courses taken in the traditional college preparatory sequence (algebra, geometry, advanced algebra, precalculus, calculus).

Given these general findings, it is important to examine the NAEP data for evidence of differential course taking by students in the race/ethnicity subgroups or by males and females. Because mathematics is required for students in grades 4 and 8 but optional for students at grade 12, different kinds of information are collected from students at different grade levels. Some data reflect students' attitudes and expectations, and other data reflect students' actual experience.

Students in grades 4 and 8 were asked to respond to the statement "If I had a choice, I would not study any more mathematics," and the strength of agreement for various groups is summarized in table 3.7. Nearly one of every six students surveyed by NAEP in grade 8 expressed a desire not to study mathematics in the future, and among Black and Hispanic students support for this view was even stronger. Some data available from other sources confirm these NAEP findings and suggest a trend toward nonenrollment in elective mathematics courses in high school, especially by minority students. A poll conducted by Louis Harris and Associates for

Table 3.7   Percent of Students Agreeing with the Statement "If I Had a Choice, I Would Not Study Any More Mathematics"

	Grade 4	Grade 8
Nation	9	16
White	8	15
Black	15	19
Hispanic	11	19
Male	11	17
Female	7	16

Source: National Center for Education Statistics. *Data Compendium for the NAEP 1992 Mathematics Assessment of the Nation and the States,* 1993

the National Action Council for Minorities in Engineering found that about 60 percent of African American and Latino students who attend public schools in grades 5–11 agreed with the statement "I will take math classes only for as long as I have to" (Leitman, Binns, and Unni 1995). Additional NAEP data suggest that too few students take elective mathematics courses in high school.

Table 3.8 contains a summary of the data obtained from twelfth-grade students' responses to a question asking how many semesters they had studied mathematics in high school, including the semester in which the survey was taken. About two-thirds of all students indicated they had studied mathematics for the equivalent of at least three years, and about one of every eight students surveyed at grade 12 had studied mathematics for less than two years. The data indicate that Black and Hispanic students are studying mathematics in high school for less time than White students. Only about one-half of the Black students surveyed had studied mathematics for three or more years, and many have studied very little mathematics at all. About one in five Black or Hispanic students indicated they had studied mathematics for less than two years; this is nearly twice as large as the proportion of White students studying so little mathematics. These differential course-taking patterns undoubtedly contribute to differential performance on the NAEP assessment at grade 12, and the relationship would likely be even stronger if the assessment at grade 12 contained more items that assessed topics in advanced high school courses.

Table 3.8 Percent of Twelfth-Grade Students Indicating Duration of Mathematics Study in High School

	Less than 4 Semesters	4 or 5 Semesters	6 or More Semesters
Nation	13	19	67
White	11	18	70
Black	20	27	51
Hispanic	19	19	60
Male	12	20	67
Female	14	18	68

Note: Row percents may not add to 100 because of rounding.

Source: National Center for Education Statistics. *Data Compendium for the NAEP 1992 Mathematics Assessment of the Nation and the States,* 1993

To understand differences in opportunity to learn important mathematical ideas, one needs to examine not only the number of semesters of mathematics studied but also which courses were taken. Table 3.9 summarizes NAEP data regarding twelfth-grade students' self-reports of having taken a particular course for at least one year.

In general, these NAEP data reflect a national trend toward increased mathematics course taking by high school students, almost certainly in response to increased graduation requirements in many states (Blank and Gruebel 1993). Given the sequential nature of high school mathematics courses and course-enrollment requirements, the data reported in tables 3.8 and 3.9 suggest that a majority of the nation's male and female high school students take three years of college preparatory mathematics. Moreover, a majority of White, Black, and Hispanic students reported studying three years of college preparatory mathematics, but the percent is smaller for Black and Hispanic students. Thus, Black and Hispanic students not only study mathematics for less time in high school than White students, but they also study less college preparatory mathematics. These differences undoubtedly contribute to the general differences in mathematical proficiency among the groups. Several other observations can be made on the basis of the data in table 3.9:

• There is essentially no difference between male and female enrollment in core, college preparatory mathematics courses in high school. There is a large difference only for courses at the extreme ends of the potential course sequence, with males more likely than females to have reported studying calculus at the high end and general mathematics at the low end.

• There is essentially no difference in the percent of White, Black, and Hispanic students who reported studying algebra 1, but there is a difference for the study of geometry, with White students somewhat more likely than Black students and far more likely than Hispanic students to report taking a course in geometry.

• The largest differences in the percent of White, Black, and Hispanic students' reported course taking are found for courses at the upper and lower ends of the potential course sequence, with White students far more likely than either Black or Hispanic students to report studying the higher-end courses (algebra 2 or precalculus) and with Black or Hispanic students far more likely than White students to report studying the lower-end courses (general math or business/consumer math).

The trajectory for high school course taking is determined to a great extent by the taking of one's first algebra course. In 1992, students in grade 12 were asked when they had first studied algebra, and about 23 percent reported having taken algebra before grade 9. In 1992, students in grade 8 were also asked to indicate the type of course in which they were enrolled, and these data are summarized in Table 3.10. One-fifth of the nation's students are studying algebra in grade 8, and more than one-fourth are

Table 3.9   Percent of Twelfth-Grade Students Indicating Taking Mathematics Course for at Least One Year

	Nation	White	Black	Hispanic	Male	Female
General Math	49	45	59	64	52	47
Consumer Math	26	25	32	29	28	24
Prealgebra	56	55	61	57	55	58
Algebra 1	87	87	84	85	86	88
Geometry	76	78	72	67	75	77
Algebra 2	61	64	49	53	60	62
Precalculus	19	21	13	12	20	19
Calculus	10	10	6	10	12	9

Table 3.10 Percent of Eighth-Grade Students Indicating Current Mathematics Course

	Algebra	Prealgebra	8th-Grade Math	Other Math
Nation	20	28	49	3
White	22	30	45	3
Black	13	23	60	4
Hispanic	12	20	62	5
Male	19	28	49	4
Female	20	28	48	3

Note: Row percents may not add to 100 because of rounding error.
Source: National Center for Education Statistics. *Data Compendium for the NAEP 1992 Mathematics Assessment of the Nation and the States*, 1993

enrolled in a prealgebra course. In contrast, the proportion of Black and Hispanic students studying either algebra or prealgebra in grade 8 is considerably lower. Overall, more than one-half of White students reported studying algebra or prealgebra in grade 8; whereas only about one-third of Black and Hispanic students are enrolled in these courses. Taken together, these data regarding mathematics in grade 8 suggest not only that the percent of students studying algebra in grade 8 did not change dramatically between 1988 and 1992 but also that non-White students remain less likely to enter high school well positioned to study advanced mathematics.

Classroom Instruction

Data available from questionnaires administered to teachers as part of the 1992 NAEP mathematics assessment provide information regarding the teaching practices and professional preparation of teachers at grades 4 and 8, although not at grade 12. Chapter 4 reviews all these data; we focus here on the data as they pertain to the demographic groups of interest in this chapter.

Instructional Support

Teachers of White students are more likely to report easy access to needed instructional resources than teachers of Black and Hispanic students. Whereas teachers of 70 percent of the White eighth-grade students reported getting all or most of the resources they needed, teachers of only 58 percent of the Black and Hispanic students reported this degree of support. A similar, but less dramatic, difference was found for teachers of fourth-grade students. But this pattern of differential access to instructional resources appears not to hold for human resources. The assistance of mathematics specialists at the fourth-grade level reportedly is available more often to teachers of a larger percent of Black students (62 percent) than to teachers of either Hispanic (50 percent) or White (50 percent) students.

Instructional Time

Instructional time is one factor that might be expected to relate to mathematical proficiency. Thus, one might expect that White students, who attain higher levels of mathematical proficiency in NAEP than Black and Hispanic students, receive more mathematics instruction. However, as the data in table 3.11 indicate, teachers of Black and Hispanic eighth-grade students actually report spending more time teaching mathematics than teachers of White students. Teachers of more than 70 percent of the White students report spending no more than four hours per week on mathematics instruction, even though the teachers of only about 63 percent of the Black and Hispanic students report this modest amount of mathematics instructional time.

Instructional time is not the only indicator of time spent studying mathematics; another factor to consider is homework. Since the teachers of White students report less instructional time, it might be expected that they compensate for this with more homework, but this is not the trend indicated by NAEP data. At the fourth-grade level, for example, teachers of Black and Hispanic students report assigning more homework than teachers of White students. The teachers of nearly three-fourths of White fourth-grade students report assigning no more than fifteen minutes of homework each day; in contrast, teachers of 58 percent of the Black students and 52 percent

Table 3.11   Percent of Eighth-Grade Students Whose Teachers Report Weekly Time for Mathematics Instruction

	Less than 2.5 Hours	Between 2.5 and 4 Hours	More than 4 Hours
Total	13	55	32
White	13	58	30
Black	14	49	38
Hispanic	12	52	36

of the Hispanic students report assigning at least thirty minutes of homework each day, and some report assigning one hour or more. At grade 8 the amounts of reported homework are similar for all students. The differences at grade 4 may be due, at least in part, to school-level or district-level policies. Responses by school administrators to questionnaires administered in conjunction with the 1992 NAEP mathematics assessment indicate that fourth-grade Black and Hispanic students are more likely than White students to attend schools in which there is a mandatory homework requirement.

Instructional Emphasis

Intergroup performance differences might also be attributed to differential opportunity to learn specific content topics, but the NAEP data indicate only a few instructional differences of this type. The reported emphasis on number and operations, measurement, geometry, and probability and statistics is similar at the eighth-grade level across the three student populations, but White eighth-grade students' teachers are more likely to report an emphasis on algebra and functions than Black and Hispanic students' teachers. At grade 4, the reported emphasis on number and operations, algebra and functions, geometry, and probability and statistics is similar, but teachers of Black and Hispanic fourth-grade students report a greater emphasis on measurement than teachers of White students.

Different instructional emphases might also be detected in the amount of attention given to various process aspects of mathematics. As in the case for content, the general picture that emerges from the NAEP data is that there are few differences. In particular, teachers of White, Black, and Hispanic students reported similar patterns of emphasis on concepts and facts, skills and procedures, and reasoning and analysis.

Instructional Media and Practices

How students gain access to mathematical ideas, along with the tools they use to explore mathematics and to solve problems, is another potential source of difference in instructional practice that could affect student proficiency. In this regard, we examined NAEP data pertaining to the classroom use of textbooks, of a variety of instructional strategies, of calculators, and of computers.

Patterns of textbook use are similar across populations in grade 4, but in grade 8, teachers of White students are more likely to report daily use of a text in contrast to teachers of Black and Hispanic students, who report less regular use of a text. At grades 4 and 8, teachers of students in all race/ethnicity groups reported similar use of small groups, objects/tools (probably manipulative materials), written reports or mathematics projects, discussions on students' solutions to problems, and student-created problems.

An area in which there were notable differences in instructional practice is the use of calculators on daily work and tests. Although teachers of fourth- and eighth-grade students reported similar access to school-owned, four-function calculators, teachers of White students at these grade levels generally reported more calculator use than teachers of Black and Hispanic students. In particular, teachers of White eighth-grade students were more likely to report daily calculator use than teachers of Black and Hispanic students (36 percent of White students in contrast to 25 percent of Black and Hispanic students). In addition, teachers of White fourth-grade students were less likely to report never using a calculator than teachers of Black and Hispanic students (49 percent of White students in contrast to 56 percent of Black and Hispanic students). Moreover, although four-function calculators were reported to be equitably distributed among students, there was somewhat less equity in access to more-sophisticated calculators. Teachers of Hispanic eighth-grade students reported more limited access to, and less instruction in the use of, scientific calculators than teachers of Black or White students.

Teachers of White students report having less and more-difficult access to computers for their students compared to the amount and ease of access reported by teachers of Black and Hispanic students. Yet, teachers of White fourth-grade students report as much actual use of computers as teachers of Hispanic students and more than teachers of Black students (about 57 percent of White and Hispanic students' teachers in contrast to 48 percent of the Black students' teachers). At grade 8, the reported frequency of use of computers is similar for teachers of all students, but White students' teachers are more likely than teachers of Black and Hispanic students to have flexible access to a computer that can be brought to class for use as needed.

Instruction and Assessment

Although most of the information presented above suggests that there are far more similarities than differences in the content of mathematics and the ways in which it is taught to White, Black, and Hispanic students in grades 4 and 8, teachers' responses to some other NAEP questions suggest some notable differences in the assessment practices of these teachers. These assessment differences may be important in understanding the differential performance of White, Black, and Hispanic students.

Overall, it appears that the frequency of assessment does not vary much among teachers of the various student populations, but the form of assessment does vary. As the data in table 3.12 indicate, teachers of Hispanic fourth-grade students are twice as likely as teachers of White fourth-grade students to report using multiple-choice tests at least once every week. Further, teachers of Black fourth-grade students are twice as likely as teach-

Table 3.12 Percent of Fourth- and Eighth-Grade Students Whose Teachers Report Their Frequency of Using Multiple-Choice Tests to Assess Student Progress

	1–2 Times per Week		1–2 Times per Month		1–2 Times per Year		Never	
	4th	8th	4th	8th	4th	8th	4th	8th
Total	6	4	43	30	24	28	27	38
White	4	4	43	28	25	29	28	40
Black	15	6	43	40	19	24	23	30
Hispanic	8	6	41	32	22	27	29	36

ers of Hispanic fourth-grade students to report this high-frequency use of multiple-choice testing. The differences at grade 8 are not as great; nevertheless, teachers of Black and Hispanic eighth-grade students are more likely to use multiple-choice testing at least once each month than are teachers of White eighth-grade students. These findings from the 1992 NAEP are consistent with the results obtained in a large-scale study of the use of standardized tests in mathematics (Madaus et al. 1992), in which it was noted that teachers who had at least 60 percent Black or Hispanic student enrollment in their classrooms were far more likely to spend classroom time using multiple-choice testing and other means of testing low-level cognitive objectives than their counterparts who had a majority of White students in their classrooms. This apparently excessive teacher attention to assessing low-level rather than high-level cognitive processes and skills most likely is related to the patterns of especially poor performance by Black and Hispanic students on complex NAEP tasks, as noted earlier in this chapter.

Differences in the use of classroom multiple-choice testing by teachers of students in the three race/ethnicity groups may be attributable, at least in part, to school-level or district-level policies. Responses by school administrators to questionnaires administered in conjunction with the 1992 NAEP mathematics assessment indicate that at grades 4 and 8, Black students are more likely than White students to attend schools in which tests are used to assign students to classes and in which tests are used for decisions regarding student retention and promotion. Moreover, at grade 4, a significantly larger percent of Black students than White students attend schools in districts in which administrators report that test scores are used for public accountability.

An alternative to multiple-choice assessment is the use of problem sets, which presumably involve students' providing not only answers but also indications of the process of solution. Teachers of White eighth-grade students are more likely to report weekly use of problem sets for assessment than teachers of Black and Hispanic eighth-grade students (more than 60 percent of the White students' teachers report this frequency of use in contrast to only about 50 percent of the Black and Hispanic students' teachers). Teachers of Black and Hispanic students were more likely to report less frequent use of problem sets as a means of student assessment (about 40 percent of the Black and Hispanic students' teachers as compared with about 30 percent of the White students' teachers). On the other hand, the pattern of differences is reversed and less pronounced at grade 4.

SUMMARY

In this chapter we have seen that the 1992 NAEP data offer some good news and some bad news. The good news is a clear indication that performance and participation differences between males and females are disappearing. Across almost every dimension of mathematics course taking and mathematics performance examined in NAEP, differences between males and females were very small and the direction of the differences was not consistent. Thus, it appears that the considerable attention given by the education community in the past decade or more to gender equity is beginning to result in a more balanced portrait of participation and achievement than found in earlier NAEP assessments. Unfortunately, the picture painted by the NAEP data regarding differences among major race/ ethnicity groups is not so positive.

The NAEP data in 1992 indicate a clear pattern of differential mathematics performance among White, Black, and Hispanic students. Although the long-term trend data indicate that the performance gap between White students and their Black and Hispanic counterparts has narrowed on basic knowledge and skills, substantial overall differences in mathematical proficiency remain. Furthermore, regarding the extremely poor performance of Black and Hispanic students relative to White students on the complex, extended tasks in the 1992 NAEP assessment, the findings provide ample reason for continued concern.

Because there was essentially no difference across the groups in students' attitudes toward mathematics, an explanation for the differential levels of performance is unlikely to reside solely within the students themselves. We have also seen that differences in several instructionally related factors, such as differences in opportunity to learn (e.g., patterns of mathematics course taking) and in the form of classroom instruction and assessment, appear related to the observed performance differences. Moreover, the 1992 NAEP findings appear to support a general conclusion that socioeconomic factors contribute heavily to the observed performance differences among White, Black, and Hispanic students. In particular, the size of differences associated with variation in socioeconomic condition was greater at each grade level and for each item type than those associated with race/ethnicity.

The NAEP data suggest that we have been quite successful in narrowing the gap between males and females in mathematics participation and performance. If this nation is to achieve the goal of the NCTM *Standards* to provide a quality mathematics education to all students, then it is clear that we will need to dedicate considerable human and financial resources to make similar progress in closing the mathematics participation and performance gaps associated with race/ethnicity and poverty.

REFERENCES

Banks, James A. *Multicultural Education: Issues and Perspectives*. Boston: Allyn and Bacon, 1989.

Blank, Rolf K., and Doreen Gruebel. *State Indicators of Science and Mathematics Education 1993*. Washington, D.C.: Council of Chief State School Officers, 1993.

Dossey, John A., Ina V. S. Mullis, Steven Gorman, and Andrew S. Latham. *How School Mathematics Functions: Perspectives from the NAEP 1990 and 1992 Assessments*. (Report No. 23-FR-02). Washington, D.C.: National Center for Education Statistics, 1994.

Dossey, John A., Ina V. S. Mullis, and Chancey O. Jones. *Can Students Do Mathematical Problem Solving? Results from Constructed-Response Questions in NAEP's 1992 Mathematics Assessment*. (Report No. 23-FR-01). Washington, D.C.: National Center for Education Statistics, 1993.

Epstein, Joyce L., and Douglas J. MacIver. *Opportunities to Learn: Effects on Eighth Graders of Curriculum Offerings and Instructional Approaches*. (Report No. 34). Baltimore, Md.: Center for Research on Effective Schooling for Disadvantaged Students, 1992.

Hart, Laurie E. "Describing the Affective Domain: Saying What We Mean." In *Affect and Mathematical Problem Solving: A New Perspective*, edited by Douglas B. McLeod and Verna M. Adams, pp. 37–44. New York: Springer-Verlag, 1989.

Johnson, Martin L. "Blacks in Mathematics: A Status Report." *Journal for Research in Mathematics Education* 15 (March 1984): 145–53.

_____. "Minority Differences in Mathematics." In *Results from the Fourth Mathematics Assessment of the National Assessment of Educational Progress*, edited by Mary M. Lindquist, pp. 135–48. Reston, Va.: National Council of Teachers of Mathematics, 1989.

Leitman, Robert, Katherine Binns, and Akhil Unni. "Uninformed Decisions: A Survey of Children and Parents about Math and Science." *NACME Research Letter* 5 (June 1995): 1–9.

Madaus, George F., Mary M. West, Maryellen C. Harmon, Richard G. Lomax, and Katherine A. Viator. *The Influence of Testing on Teaching Math and Science in Grades 4–12* (Report of Grant No. SPA8954579 funded by the National Science Foundation). Boston: The Center for the Study of Testing, Evaluation and Educational Policy, Boston College, 1992.

Matthews, Westina. "Influences on the Learning and Participation of Minorities in Mathematics." *Journal for Research in Mathematics Education* 15 (March 1984): 84–95.

Matthews, Westina, Thomas P. Carpenter, Mary M. Lindquist, and Edward A. Silver. "The Third National Assessment: Minorities and Mathematics." *Journal for Research in Mathematics Education* 15 (March 1984): 165–71.

Meyer, Margaret R. "Gender Differences in Mathematics." In *Results from the Fourth Mathematics Assessment of the National Assessment of Educational Progress,* edited by Mary M. Lindquist, pp. 149–59. Reston, Va.: National Council of Teachers of Mathematics, 1989.

Mullis, Ina V. S., John A. Dossey, Jay R. Campbell, Claudia A. Gentile, Christine O'Sullivan, and Andrew S. Latham. *NAEP 1992 Trends in Academic Progress.* (Report No. 23-TR-01). Washington, D.C.: National Center for Education Statistics, 1994.

Naifeh, Mary, and Sharif Shakrani. *Math Matters: The Relationship Between High School Mathematics Course-Taking and Proficiency on the NAEP Assessment.* Washington, D.C.: National Center for Education Statistics, 1996.

National Center for Education Statistics. *Data Compendium for the NAEP 1992 Mathematics Assessment of the Nation and the States.* (Report No. 23-ST-04). Washington, D.C.: National Center for Education Statistics, 1993.

Oakes, Jeannie. *Multiplying Inequalities: The Effects of Race, Social Class, and Tracking on Opportunities to Learn Mathematics and Science.* Santa Monica, Calif.: The Rand Corporation, 1990a.

————."Opportunities, Achievement, and Choice: Women and Minority Students in Science and Mathematics." In *Review of Research in Education,* Vol. 16, edited by Courtney B. Cazden, pp. 152–222. Washington, D.C.: American Educational Research Association, 1990b.

Reyes, Laurie H. "Attitudes and Mathematics." In *Selected Issues in Mathematics Education,* edited by Mary M. Lindquist, pp. 161–84. Evanston, Ill.: National Society for the Study of Education and National Council of Teachers of Mathematics, 1980.

Reyes, Laurie H., and George M. A. Stanic. "Race, Sex, Socioeconomic Status, and Mathematics." *Journal for Research in Mathematics Education* 19 (January 1988): 26–43.

Spencer, Bruce D. "A Study of Eligibility Exclusions and Sampling: 1992 Trial State Assessment." In *Studies of The National Academy of Education Panel on the Evaluation of the NAEP Trial State Assessment: 1992 Trial State Assessment,* pp. 1–67. Stanford, Calif.: National Academy of Education, 1994.

Stanic, George M. A., and Laurie E. Hart. "Attitudes, Persistence, and Mathematics Achievement: Qualifying Race and Sex Differences." In *New Directions for Equity in Mathematics Education,* edited by Walter G. Secada, Elizabeth Fennema, and Lisa Byrd Adajian, pp. 258–76. New York: Cambridge University Press, 1995.

Welch, Wayne W., Ronald E. Anderson, and Linda J. Harris. "The Effects of Schooling on Mathematics Achievement." *American Educational Research Journal* 19 (Spring 1982): 145–53.

4

NAEP Findings regarding the Preparation and Classroom Practices of Mathematics Teachers

Mary Montgomery Lindquist

THE *PROFESSIONAL STANDARDS for Teaching Mathematics* (NCTM 1991) sets forth standards for the professional development of teachers of mathematics. The *Standards* expresses a vision of well-prepared teachers who know and understand mathematics, how students learn mathematics, and how to orchestrate that learning for all students. Information gathered from the teacher questionnaires of the National Assessment of Educational Progress (NAEP) gives us a glimpse into the opportunities that teachers have had for professional development as well as into some of their classroom practices and beliefs.

Although this chapter focuses on information from the NAEP teacher questionnaires, it is supplemented by other sources. Data from the NAEP student questionnaires are cited, especially when students' perceptions of classroom practice differ from those of teachers. Because relatively few twelfth-grade students participating in the assessment were enrolled in mathematics classes, there were no NAEP teacher questionnaires given to high school teachers. To augment the discussion of teachers at all three levels, the Council of Chief State School Officers (CCSSO) report, *State Indicators of Science and Mathematics Education 1995* (Blank and Gruebel 1995), is drawn on for information about high school mathematics teachers. This chapter also uses data from a 1993 national survey (Weiss 1994) regarding background, practices, and beliefs of mathematics teachers of grades 1–4, grades 5–8, and grades 9–12.

The background information from NAEP teachers' questionnaires as well as from the other questionnaires (see chapter 1 for a discussion of the NAEP questionnaires) can be linked to performance for students in grades 4 and 8. This chapter includes student performance data when this background information provides some insights into student achievement. However, care must be taken not to make inferences of a causal nature but to use the information only for further probing or reflecting on patterns of responses.

WHO ARE OUR TEACHERS?

The data from NAEP, supplemented by the documents mentioned in the introduction, paint a broad-brush portrait of mathematics teachers. This composite picture gives a glimpse of some demographics and the educational backgrounds of mathematics teachers. In particular, there is information about amount of teaching experience, gender and race/ethnicity, level and type of teacher certification, and major. In addition to stating what mathematics and pedagogical courses they had taken, teachers were

HIGHLIGHTS

- Students are taught predominantly by White females with 10 to 25 years of teaching experience, although with somewhat less experience teaching mathematics. Many teachers lack recent updating of their background in mathematics and its teaching.
- Mathematics teachers perceive themselves as being adequately or well prepared to teach mathematics in the grade levels for which they are certified. Almost two-thirds of students in grade 8 are taught by teachers who have an undergraduate degree in mathematics or mathematics education.
- The time allotted to mathematics instruction decreases from grade 4 to grade 8 as the time spent on homework increases. There appears to be little performance difference directly attributable to the length of the class period, which suggests the importance of quality instruction and high expectations of students.
- Although there is some variety in instructional activities, classroom instruction still relies on lecture and textbooks. Communication—especially writing—has yet to play a critical role in mathematics instruction. The amount of small-group work depends upon the viewpoint of the responder: students do not perceive participating in as much small-group work as teachers say occurs.
- Teachers have some, but not all, of the resources they think they need to teach mathematics effectively.
- Assessment occurs often but mainly through pencil-and-paper testing. The seven weeks of school consumed by testing is unsubstantiated by data. While we know the frequency of testing, we do not know the amount of time associated with each of the "tests."

asked to reflect on their preparation to deal with specific mathematical topics as well as particular teaching strategies. Each of these areas is discussed in the sections that follow.

Years of Teaching

The number of years of teaching gives an idea about the recency of the initial preparation of teachers. The NAEP data indicate that there is little difference in the numbers of years of teaching between teachers of students in grades 4 and 8. Approximately half of the students at both grades have teachers who have taught from ten to twenty-five years, and another 18 percent of the students at each grade level have teachers who have taught twenty-five or more years. As shown in table 4.1, teachers of eighth-grade students have had more overall teaching experience than mathematics teaching experience. For example, 44 percent of students in grade 8 have teachers who have taught mathematics ten or fewer years, while only 35 percent have teachers with ten or fewer years of overall experience. Although the NAEP data do not explain the difference between the number of years of overall versus mathematics teaching experience, several plausible explanations can be given. The difference may be partially attributed to teachers certified in other fields who added mathematics to their certification. This often happens at times, such as in the 1980s, when there is a shortage of mathematics teachers. Possibly, K–8 generalists who may have been teaching subjects other than mathematics are now teaching mathematics. The cause of the difference may not be as important as the realization that eighth-grade students have teachers of mathematics who come from many different preparations and experiences in teaching mathematics.

Table 4.1 Eighth-Grade Teachers' Reports on Number of Years of Overall Teaching Experience and Mathematics Teaching Experience

	10 Years or Less		More than 10 Years but Less than 25 Years		25 Years or More	
	Percent of Students	Average Proficiency	Percent of Students	Average Proficiency	Percent of Students	Average Proficiency
Overall	35	265 (1.7)	47	269 (1.3)	18	275 (2.2)>
Mathematics	44	265 (1.4)	43	271 (1.5)	13	276 (2.9)>

> The value for 25 years or more is significantly higher than the value for 10 years or less at about the 95 percent confidence level.

Note: The standard errors of the estimated proficiencies appear in parentheses.

Source: National Center for Education Statistics, *Data Compendium for the NAEP 1992 Mathematics Assessment of the Nation and the States,* 1993

The data in table 4.1 also show that years of mathematics teaching experience may have had an impact on student performance levels on NAEP. For example, students who had teachers with twenty-five or more years of mathematics teaching experience had a significantly higher average NAEP scale score than students who had teachers with ten or fewer years of mathematics teaching experience. Although there are significant differences in the proficiency of students who had teachers with the least experience—compared with those who had teachers with the most experience—this may not be accounted for by experience alone. For example, teachers with seniority are often assigned the classes with better-prepared students.

There is no significant difference in years of teaching experience with regard to gender or race/ethnicity. For example, about the same percentage of male and female students were taught by teachers with ten or fewer years of experience and about the same percentage of White, Black, and Hispanic students were taught by teachers with twenty-five or more years of experience. Chapter 3 provides more information on data from the teacher questionnaires as they pertain to race/ethnicity subgroups.

Although the NAEP data provide no information about the longevity of high school teachers, the CCSSO survey (Blank and Gruebel 1995) reports the percent of teachers under age 30 and over age 50. About 15 percent of the nation's high school mathematics teachers are under 30 and about 23 percent are over 50, as calculated on the basis of the thirty-three states that responded. These data point to the need for ongoing professional development opportunities. For example, most high school mathematics teachers finished college long before the advent of graphing calculators.

Race/Ethnicity and Gender

In contrast to the lack of differences of various subpopulations in terms of longevity of teaching, the data in table 4.2 show great differences in the percents of students taught by someone of the same gender and from the same race/ethnicity subgroup. About seven-eighths of fourth-grade students and about three-fifths of eighth-grade students, regardless of their gender or race/ethnicity, are taught by female teachers. Only about one-fourth of Black students in grades 4 and 8 have Black teachers, and little more than one-tenth of the Hispanic students in grade 4 and one-fifth in grade 8 have Hispanic teachers. Very few White students have Black or Hispanic teachers.

Additional data from NAEP indicate that there was an increase from 1990 to 1992 in the percent of students in grade 4 who were taught by White teachers: 84 percent in 1990 and 90 percent in 1992. At grade 4 between those two years, there was a striking decline in the proportion of Black students taught by Black teachers: from 50 percent in 1990 to 26 percent in

Table 4.2 Teachers' Reports on Their Gender and Race/Ethnicity

	Teachers' Gender		Teachers' Race/Ethnicity		
	Male	Female	White	Black	Hispanic
Grade 4	**Percent of Students**		**Percent of Students**		
White	12	88	95	4	1
Black	13	87	72	26	0
Hispanic	13	87	76	9	13
Male	13	87	89	8	2
Female	12	88	90	7	2
Grade 8	**Percent of Students**		**Percent of Students**		
White	41	59	95	2	1
Black	36	64	73	25	1
Hispanic	42	58	68	10	20
Male	41	59	89	7	3
Female	39	61	89	6	3

Source: National Center for Education Statistics, *Data Compendium for the NAEP 1992 Mathematics Assessment of the Nation and the States*, 1993

1992. No similar change occurred from 1990 to 1992 in grade 8; in both years about 90 percent of White students were taught by White teachers and about 25 percent of Black students were taught by Black teachers. As we strive to open the doors of mathematics to all students, we need a teaching force that reflects their diversity.

The CCSSO study (Blank and Gruebel 1995) affords some information about the gender and race/ethnicity of high school mathematics teachers. About half of the high school mathematics teachers are male, but the percentage varies greatly from state to state. In general, the southeastern states have more female than male mathematics teachers, whereas the opposite is true in the midwestern states. From 1990 to 1994, there was an increase in females teaching high school mathematics. Similar to the situation in elementary and middle schools, there is also a recognizable need for mathematics teachers from diverse racial and ethnic backgrounds at the high school level. According to the CCSSO report, at the secondary level about 33 percent of the students and 14 percent of the mathematics teachers are members of a minority. Although the CCSSO report does not link race/ethnicity of students with teachers, the percents themselves point to an imbalance. States with a greater percentage of minority students tend to have a greater percentage of minority teachers; however, in every state except Hawaii, the proportion of minority mathematics teachers to that of minority students is small.

Level and Type of Certification

NAEP data show that less than one-tenth of students in grades 4 and 8 were taught by teachers who hold a temporary, provisional, probational, or emergency certificate or who were not certified. About one-third of students in these grades were taught by teachers who hold a regular certification, and the rest (teachers of 56 percent of fourth-grade students and 60 percent of eighth-grade students) were taught by teachers holding the highest level of certification, that is, a permanent or long-term certification.

Teachers may be certified in mathematics, in general education, or otherwise. The majority of fourth-grade students were taught by teachers holding an education certification, although about one-tenth of students at this grade were taught by teachers with a middle school or secondary mathematics certification. In contrast, about three-fourths of the students in grade 8 had teachers with a middle school or secondary mathematics certification. There is a slight, but not significant, decline from 1990 to 1992 in the percentage of eighth-grade students taught by teachers with mathematics certification. It will be interesting to watch the shifts in these percentages in the coming years as the middle school movement gains momentum.

The CCSSO report also provides information on the certification of high school teachers. Results from that report reveal that about 90 percent of the mathematics teachers were certified in secondary mathematics (Blank and Gruebel 1995).

Type of Degree and Majors

Given that certification is a measure only of initial preparation, it is also useful to know the type of degrees that teachers have obtained. NAEP data indicate that about 55 percent of students in grades 4 and 8 were taught by teachers holding a bachelor's degree. The remaining 45 percent were taught by those with a master's or specialist's degree; there were very few teachers at these grade levels with doctorates.

The type of degree gives additional information about the preparation of teachers of mathematics. Table 4.3 shows the percents of students with teachers who majored in mathematics, mathematics education, education, or other areas. About 80 percent of students in grade 4 were taught by teachers who majored in education at the undergraduate level, and a similar percent of students at that grade level were taught by teachers with graduate degrees in education. In grade 8, about 43 percent of students had teachers who majored in mathematics and 15 percent had teachers who majored in mathematics education. Another 29 percent were taught by teachers who majored in education, and the remainder of students were taught by teachers with various other majors. Clearly, both the initial college mathematics preparation and the graduate school education of teachers of eighth-grade students are more substantial than those of teachers of fourth-grade students. The difference between fourth- and eighth-grade students whose teachers have graduate degrees in mathematics is not as pronounced as that for students whose teachers have undergraduate degrees in mathematics.

Table 4.3 Teachers' Reports on Their Undergraduate and Graduate Majors

	Undergraduate Major		Graduate Major	
	Grade 4	Grade 8	Grade 4	Grade 8
Mathematics	6	43	2	21
Mathematics Education	2	15	4	19
Education	81	29	81	47
Other	12	13	13	13

Note: The data in the table are expressed as the percent of students whose teachers responded in each of the various categories.
Source: National Center for Education Statistics, *Data Compendium for the NAEP 1992 Mathematics Assessment of the Nation and the States,* 1993

Further examination of the data reveals few differences in performance for students in grade 4 who have teachers with different undergraduate or graduate majors. In seven of the undergraduate and graduate majors shown in table 4.3, the proficiency level of fourth-grade students ranged from 217 to 220 on the NAEP scale. The one exception was the performance of those fourth-grade students who had teachers with undergraduate degrees in mathematics education. Although their average proficiency was 231, only 2 percent of the students had teachers with this major. It is an interesting category to watch, especially if opportunities increase for elementary teachers to specialize in mathematics education.

At grade 8, students with teachers who had majored in mathematics or mathematics education had an average proficiency of about 270, while the average proficiency of students whose teachers majored in education or some other field was about 260. (The one exception to this involved eighth-grade students who had teachers with graduate degrees in education, but those teachers may have had an undergraduate degree in mathematics.) Thus, there is some indication that the major of teachers of students in grade 8 makes more difference with respect to student performance on the NAEP test than it did for students in grade 4. There are other recent studies that address the relationship between teachers' majors and student performance. In her doctoral dissertation, Bell (1992) found that scores on the South Carolina state test for sixth-grade students were not significantly higher for those students taught by mathematics-certified teachers than for those taught by elementary-certified teachers. Eighth-grade scores were significantly higher, however, for those students taught by mathematics-certified teachers than for those taught by teachers not certified in mathematics.

Almost 70 percent of the teachers (Blank and Gruebel 1995) whose main assignment is teaching mathematics in grades 9–12 had an undergraduate major in mathematics. This percent is about 60 percent if teachers of grades 7 and 8 are included, but the percent varies greatly from state to state: from a low of 45 percent to a high of about 80 percent. Also, there is a great variation from state to state in the percentage of grade 7–12 teachers who teach some mathematics but whose main assignment is not mathematics. In general, the mathematics background of these teachers is not as strong. Overall, about 59 percent of all teachers who teach mathematics in grades 7–12, either as their main assignment or a part-time assignment, have a major or minor in mathematics or mathematics education. This discrepancy begs for further study of the performance of students whose teachers have this varied preparation.

Mathematical Background

In 1991, the Mathematical Association of America (MAA) set forth standards for the mathematics preparation of teachers (Leitzel 1991). The content standards specific to the preparation of K–4 teachers include number and use of number; geometry and measurement; patterns and functions; and collecting, representing, and interpreting data. The content standards for teachers in grades 5–8 extend each of the K–4 standards and place greater emphasis on algebra, probability and statistics, and concepts of calculus. The NAEP data in table 4.4 show the percents of students in grades 4 and 8 whose teachers have studied particular mathematical content areas, either in college or through in-service courses.

The study of number systems and numeration is often included in college courses for elementary teachers. Almost 90 percent of students at each grade level had teachers who have studied this topic in college. Number

Table 4.4 Teachers' Preparation in Mathematics Content Areas

	One or More College Courses	Part of a College Course	In-Service Training	Little or No Exposure
Number Systems and Numeration				
Grade 4	53	36	34	6
Grade 8	56	32	21	8
Measurements in Mathematics				
Grade 4	43	39	34	10
Grade 8	47	34	28	9
Geometry				
Grade 4	33	39	27	17
Grade 8	66	23	25	7
Probability and Statistics				
Grade 4	45	23	20	24
Grade 8	67	22	18	8
Abstract/Linear Algebra				
Grade 4	31	21	7	46
Grade 8	67	15	11	15
Calculus				
Grade 4	13	7	1	78
Grade 8	66	7	2	25

Note: The data in the table are expressed as the percent of students whose teachers responded in each of the various categories.

Source: National Center for Education Statistics, *Data Compendium for the NAEP 1992 Mathematics Assessment of the Nation and the States*, 1993

and numeration is a slightly more popular topic for in-service courses for fourth-grade teachers than for eighth-grade teachers, which reflects the difference between elementary and middle school curricula.

About 80 percent of the students in each grade had teachers who have taken one or more college courses in measurement or who have studied topics in measurement as a part of a course. About the same percentage of students in grades 4 and 8 had teachers who have studied measurement in some in-service course. The study of geometry in one or more college courses was about twice as common for teachers of students in grade 8 (66 percent) as for teachers of students in grade 4 (33 percent). The same is true of abstract/linear algebra: 31 percent for teachers of fourth-grade students and 67 percent for teachers of eighth-grade students. Because calculus is a prerequisite to more-advanced algebra courses such as linear algebra and abstract algebra—and only 13 percent of fourth-grade students had teachers with calculus in their repertoire—the fourth-grade response more realistically refers to beginning college algebra rather than abstract/linear algebra. About two-thirds of students in grade 4 and about nine-tenths of students in grade 8 had teachers who reported having studied probability and statistics. The large percentage of teachers who reported having studied probability and statistics may include those who have taken educational statistics courses, which suggests that teachers may not have explored probability and statistics from a mathematical or pedagogical viewpoint.

The Weiss (1994) study gives us insight about how well the teachers of grades 1–4 and grades 5–8 rate their preparation to teach the different strands of mathematics. About half of the teachers in grades 1–4 feel very well qualified and the other half feel adequately qualified to teach whole numbers, fractions, decimals, and measurement. They also feel that they are well qualified to teach patterns and geometry. They do not, however, feel qualified to teach algebra, probability and statistics, or topics traditionally considered to be more secondary than elementary. Middle school teachers report more confidence in their qualifications to teach algebra, functions, and probability and statistics than do elementary teachers but feel less qualified to teach such topics as discrete mathematics or introductory calculus than do high school teachers.

The amount of emphasis teachers report placing on different topics in their teaching gives an indication of the opportunity that students have to learn these topics. Teachers were asked by NAEP about the degree of emphasis (heavy, moderate, or little) that they place on the following topics: number, measurement, geometry, data, and algebra. About 92 percent of students in grade 4 and about 76 percent of students in grade 8 have teachers who heavily emphasize number work. The only other topic for which teachers reported heavy emphasis was algebra, but this was only for advanced students who probably were taking algebra in grade 8. According to NAEP data, topics in measurement and geometry were given moderate emphasis. In 1990, 70 percent of fourth-grade students and 50 percent of eighth-grade students had teachers who reported placing moderate emphasis on measurement, and these figures rose in 1992 to 81 percent in grade 4 and 69 percent in grade 8. From 1990 to 1992, the percent of students whose teachers reported moderate emphasis on geometry rose from 58 percent to 71 percent in grade 4 and from 51 percent to 71 percent in grade 8. In 1992, more than half of fourth-grade students had teachers who placed little emphasis on data analysis (including statistics and probability) and two-thirds had teachers who placed little emphasis on algebra. About 30 percent of eighth-grade students had teachers who place little or no emphasis on data analysis, and only about 12 percent had teachers who place little or no emphasis on algebra. Between grade levels, then, one can see increasing percentages of students whose teachers see the need to emphasize topics other than number. The increase implies a need for strong teacher preparation in these other mathematics topics.

Pedagogical Background

Professional Standards for Teaching Mathematics (NCTM 1991) emphasizes the need for teacher knowledge about students as learners of mathematics as well as mathematics pedagogy. The NAEP questionnaires gathered information about the opportunities teachers have to take courses that are likely to address these topics. About 92 percent of students in grade 4 had teachers who have had at least one college course in methods of teaching elementary school mathematics, and the remainder have had at least some college preparation. About 50 percent of students in this grade level had teachers who reported having had some in-service work focused on teaching mathematics. In grade 8, only 58 percent of the students had teachers who reported having had a course in middle school mathematics methods. Middle school is a relatively new area for special programs in colleges and universities, and teachers who completed their academic degrees more than ten years ago may not have had the opportunity to be involved in these programs. These teachers may, however, have had training in elementary or secondary mathematics methods.

The development of teachers does not stop with the initial collegiate experience. With the changing population, expectations for students, and higher standards, teachers have a constant need for further study. The NAEP data indicate that about half of the students in grades 4 and 8 have teachers who have had some in-service work in the teaching of mathematics. Nationwide, there was a significant increase from 1990 to 1992 in the amount of in-service training that exceeded sixteen hours. In grade 8, there was an increase in the percent of students whose teachers had more than sixteen hours and a corresponding decrease in the percent of students whose teachers had participated in one to fifteen hours of in-service training. The increase in time spent in in-service experiences is consonant with the call for ongoing and substantive experiences rather than short, one-day remedies. There is little doubt that programs such as the Dwight D. Eisenhower program have increased the opportunities for teachers to participate in mathematics in-service programs.

The Weiss (1994) study examined when teachers took their last college course in mathematics and in mathematics education as well as the amount of time spent in mathematics in-service training. About half of the teachers in grades 1–4 and in grades 5–8 and about two-fifths of the high school teachers had not taken a mathematics course in the last ten years. About two-fifths of all teachers had not taken a mathematics education course in the prior ten years. In the previous three years, about two-fifths of the early childhood and middle school teachers had less than six hours of in-service work in mathematics or mathematics education. In interpreting the middle school data, it must be noted that many middle school teachers may not be teaching mathematics. At the high school level, about one-fourth of the teachers have had fewer than six hours while over half have had more than sixteen hours of in-service training in mathematics or mathematics education. This information, along with that from the NAEP questionnaires, indicates that teachers still do not have the ongoing, substantial in-service experiences that they may need to be current with the rapid changes of today.

Teachers were asked in the NAEP questionnaires about their college or in-service experience in regard to a variety of topics whose emphasis has recently increased in the intended curriculum or instruction. Table 4.5 shows the percent of students in grades 4 and 8 whose teachers had college or in-service experiences in these topics. Estimation, a skill we use daily, only recently has been acknowledged as an important topic for elementary and middle school curricula. However, almost 80 percent of the students in grades 4 and 8 had teachers who have had college or in-service experiences with estimation. Problem solving and using manipulatives have been a focus of mathematics curriculum and instruction since the early 1980s and have been reinforced in the NCTM (1989) *Curriculum and Evaluation Standards for School Mathematics*. The proportion of students—more than 90 percent in both grade levels—whose teachers have had opportunities in both of these areas reflects the extended period of emphasis. In 1992, about 60 percent of fourth-grade students and about 70 percent of eighth-grade students had teachers who reported their students had access to school-owned calculators. It is interesting to note how closely these percents reflect the percents of students whose teachers have had college or in-service courses on using calculators. Recently, there has also been more emphasis

Table 4.5 Teachers' Reports on College or In-Service Courses on Specific Topics in Mathematics and in Special Areas

	Grade 4	Grade 8
Estimation	80	78
Problem Solving	92	93
Use of Manipulatives	93	88
Use of Calculators	59	76
Understanding Students' Thinking about Mathematics	72	66
Gender Issues	32	41
Teaching Students from Different Cultural Backgrounds	42	50

Note: The data in the table are expressed as the percent of students whose teachers have studied these topics.

Source: National Center for Education Statistics, *Data Compendium for the NAEP 1992 Mathematics Assessment of the Nation and the States*, 1993

on understanding what students are thinking in mathematics. It appears that this has also been given emphasis in college or in-service courses, since 72 percent of fourth-grade students and 66 percent of eighth-grade students have teachers who reported studying this topic.

A goal of the NCTM *Curriculum and Evaluation Standards* is for every student to develop the mathematical power needed for today's world. Furthermore, the *Standards* documents recognize the need for teachers to be aware of gender and cultural differences in order to reach all students. The data in table 4.5 concerning gender issues and teaching students from different cultural backgrounds indicate that some attention has been given to these topics, but not all students have been afforded the opportunity of having teachers with training in these areas. The Weiss (1994) study asked specifically about teachers' perceptions of their preparation to work with several different groups. In general, teachers felt least prepared to work with students with limited English proficiency and with learning disabilities.

Although NAEP data give a broad overview of the opportunities that teachers have had in areas such as coursework in mathematics content and pedagogy, the data say little about the quality of the experiences or their effect on classroom teaching. There is, however, some information from the NAEP questionnaires about teachers' views of their preparedness to teach mathematical concepts and procedures as well as to use calculators and computers to teach mathematics. The data in table 4.6 indicate that students in grade 4 have teachers who perceived themselves to be slightly less prepared in all three of these areas than the teachers of students in grade 8. For example, only 34 percent of fourth-grade students have teachers who indicate they are well prepared for using calculators in the classroom, while 58 percent of eighth-grade students have teachers who feel well prepared. However, if the categories of Very Well Prepared and Moderately Well Prepared are combined, there is little difference between the two grade levels. In summary, students in both grades 4 and 8 have teachers who feel very well prepared to teach concepts and procedures, less prepared to deal with calculators, and much less prepared to use computers in their teaching. The data clearly point to the need for additional opportunities for teachers to work with technology and to prepare to use technology in teaching mathematics.

Although there is no information about high school mathematics teachers in the NAEP data, the Weiss (1994) study provides comparative data

Table 4.6 Teachers' Reports on Degree of Preparation for Teaching Mathematics Concepts and for Using Computers and Calculators in the Classroom

	Very Well Prepared	Moderately Well Prepared	Not Very Prepared	Not at All Prepared
Mathematics Concepts and Procedures				
Grade 4	85	15	0	0
Grade 8	93	7	0	0
Computers				
Grade 4	15	40	34	10
Grade 8	21	34	32	13
Calculators				
Grade 4	34	47	15	3
Grade 8	58	35	7	1

Note: The data in the table are expressed as the percent of students whose teachers responded in each of the various categories.

Source: National Center for Education Statistics, *Data Compendium for the NAEP 1992 Mathematics Assessment of the Nation and the States*, 1993

about their preparation to use various instructional strategies. When those data are examined, it is clear that secondary school teachers do not feel prepared to deal with as many different instructional strategies as their elementary and middle school counterparts do. This is especially evident, but not surprising, in terms of using manipulatives, teaching heterogeneous groups, and integrating mathematics with other subject areas. Findings from the Weiss study reconfirm the need for more teacher preparation in using computers in teaching mathematics. Secondary teachers, however, do indicate that they are prepared to use calculators as an integral part of teaching mathematics—a contrast to the perception of elementary teachers.

WHAT HAPPENS IN CLASSROOMS?

We have just examined the background of teachers and their opportunities for professional development. This part of the chapter examines data related to instructional conditions, instructional activities, resources, and assessment. Although no analyses were done to relate background and opportunities of teachers with classroom practices, it is important to know what teachers and students perceive is happening in the classrooms.

Instructional Conditions

There are many variables that could be classified as instructional conditions. In particular, NAEP examined the amount of time spent on mathematics, the amount of homework, and class size. These variables are discussed here in general; chapter 3 examines differences in these conditions among various subgroups.

The time devoted to a discipline is a measure of how that discipline is valued at each level of schooling. Traditionally, in elementary schools, reading is a priority, followed by mathematics. As shown by the data in table 4.7, almost three-fourths of the fourth-grade students spend four or more hours in mathematics class each week. There is a decline from grade 4 to grade 8 in the percent of students receiving four or more hours' instruction each week in mathematics. Since most eighth-grade students spend fewer than four hours a week in mathematics, there may be a problem with offering courses such as algebra in grade 8. Having less time for algebra in grade 8 than in high school may not present difficulties for the most capable students, but if a school or district opts to have a large percent of students take algebra in grade 8, consideration must be given as to whether sufficient class time is allotted.

A further analysis of the NAEP data (Dossey et al. 1994) indicates that less time was spent in mathematics by more students in the top one-third

Table 4.7 Teachers' Reports on Amount of Time Spent on Mathematics Instruction Each Week

	Two and One-Half Hours or Less		More than Two and One-Half Hours, but Less than Four Hours		Four Hours or More	
	Percent of Students	Average Proficiency	Percent of Students	Average Proficiency	Percent of Students	Average Proficiency
Grade 4	5	223	24	223	71	216
Grade 8	13	269	55	270	32	267

Source: National Center for Education Statistics, *Data Compendium for the NAEP 1992 Mathematics Assessment of the Nation and the States,* 1993

of the schools than by those in the bottom one-third of the schools. (The designation of the top and bottom one-third of the schools is based on a sort of schools by their students' average performance on NAEP.) In the top one-third of the schools, 60 percent of the students in grade 4 and 29 percent of the students in grade 8 had four or more hours in mathematics. In the bottom one-third of the schools the corresponding percents were 82 percent and 37 percent. That is, students in the bottom one-third of the schools spend more time in mathematics classes than their counterparts in the top one-third of schools. This analysis may help explain why there is little difference in performance associated with the length of time spent on mathematics, as shown by the average proficiency data in table 4.7. These NAEP results suggest that time alone is not the answer; the quality of instruction and the expectations for students also must be considered.

Although students in grade 8 spend less time in mathematics classes, they do spend more time on homework than students in grade 4. The data in table 4.8 show that about half the students in grade 4 spent fifteen minutes a day on mathematics homework. In contrast, about half the students in grade 8 spent thirty minutes. There appears to be little difference in performance for students in grade 4 relative to the amount of homework. In grade 8, however, there is a significant difference between the performance of students whose teachers report they give no homework and that of those who had some homework (fifteen or more minutes). Again, an argument could be raised in favor of expectations and holding students to high standards.

One might expect that class size would have an effect on performance, but the overall NAEP data show no pattern of effect of class size on performance. For example, in grade 4, 20 percent of the students are in classes with one to twenty students, 36 percent in classes with twenty-one to twenty-five students, and 43 percent in classes with forty-three students.

Table 4.8 Teachers' Reports on the Amount of Homework Assigned Daily

	Grade 4		Grade 8	
	Percent of Students	**Average Proficiency**	**Percent of Students**	**Average Proficiency**
None	6	221 (2.4)	3	238 (5.1)
15 Minutes	52	220 (1.3)	28	263 (1.7)
30 Minutes	37	217 (1.6)	49	268 (1.4)
45 Minutes	4	201 (4.8)	16	282 (3.4)
An Hour or More	1	206 (11.6)	4	287 (5.0)

Source: National Center for Education Statistics, *Data Compendium for the NAEP 1992 Mathematics Assessment of the Nation and the States*, 1993

Note: The standard errors of the estimated proficiencies appear in parentheses.

Their average proficiency levels, respectively, are 220, 216, and 220. More-refined analyses of NAEP data (Dossey et al. 1994) show some relation-ships between class size and performance. For example, in both grades 4 and 8, more students in the bottom one-third of the schools were in larger classes than the corresponding students in the top one-third of the schools, a finding consistent with that reported by Glass et al. (1982).

Instructional Activities

In addition to the amount of time devoted to a subject and the number of students in a class, the instructional activities in which students partici-pate should influence their performance. Today, there is a call for the active engagement of students in doing and communicating about mathematics. From both the NAEP data and the Weiss survey, we gain information about teachers' perceptions of how these activities occur in their classrooms.

It is apparent that mathematics instruction is dependent on textbooks. The NAEP data in table 4.9 show that more than three-fourths of the stu-dents in each of grades 4 and 8 use textbooks daily. Teachers also report that they cover more than three-fourths of their mathematics textbook, and about the same number of teachers rate their texts as good to excellent (Weiss 1994). If the textbooks provide problem sets of high quality, then they certainly may ease the life of busy teachers. Because of this reliance on textbooks, it is incumbent on the mathematics community to see that text-books reflect the type of mathematics in which we expect students to en-gage.

Table 4.9 Teachers' Reports on Frequency with Which Students Use Textbooks and Worksheets

	Percent of Students		
	Almost Every Day	At Least Once a Week	Less than Weekly
Do Mathematics Problems from Textbooks			
Grade 4	76	20	4
Grade 8	83	14	3
Do Mathematics Problems from Worksheets			
Grade 4	26	56	18
Grade 8	12	52	36

Source: National Center for Education Statistics, *Data Compendium for the NAEP 1992 Mathematics Assessment of the Nation and the States*, 1993

The data in table 4.9 also reveal that the practice of using problems from worksheets is not as common as the use of problems from textbooks; however, more than half of the students have weekly experience with problems from a source other than the textbook. Although there is little indication of the type of problems provided by the worksheets, the data show that teachers are willing to supplement the textbook material.

The question naturally arises about how teachers are presenting textbook material. In the study by Weiss (1994), 63 percent of the teachers at grades 1 through 4 said they never had "students listen and take notes" as they lectured, whereas 73 percent of the teachers at grades 9 through 12 said this was a daily occurrence. Elementary teachers most likely were responding to the "take notes" part of the conjunctive statement because lecturing was the primary mode of instruction at all levels. When asked what activities took place in their most recent mathematics class, 82, 90, and 94 percent, respectively, of the teachers of grades 1–4, 5–8, and 9–12 indicated lecture.

For too long, mathematics has been a silent subject in our schools: students watch the teacher present, and then they do problems from the textbooks. The call to do more communicating, both written and oral, is a call to help students develop an understanding about mathematics and a way to use mathematics. Although mathematics is often considered a language, for many students it is only a language of symbols that have little connection with ordinary language or applications. Thus, teachers are being encouraged to have students talk more about mathematics as they explain their work or justify their conjectures. The data in table 4.10 look only at writing as a means of communicating and only in terms of problem solving. These data are for all students in the fourth-grade and eighth-grade NAEP sample and for only those students in the twelfth-grade sample who reported current enrollment in a mathematics course. It is evident from these data that students are not accustomed to writing in mathematics (in connection with describing how they solved problems), nor do they create problems to be solved. About one-fifth of the students in grades 4 and 8 and fewer in grade 12 have weekly experiences in writing about how they solved a problem. Thus, it is not surprising that the constructed-response

Table 4.10 Teachers' and Students' Reports on Frequency of Writing and Posing Problems

	At Least Weekly		Less than Once a Week		Never or Hardly Ever	
	Teachers' Reports[a]	Students' Reports	Teachers' Reports[a]	Students' Reports	Teachers' Reports[a]	Students' Reports
Write a few sentences about how you solved a mathematical problem						
Grade 4	19	—	36	—	45	—
Grade 8	21	21	38	18	41	62
Grade 12 (taking math)	—	15	—	17	—	68
Make up mathematics problems for other students to solve						
Grade 4	21	—	47	—	31	—
Grade 8	8	8	34	15	59	77
Grade 12 (taking math)	—	5	—	9	—	86

[a]These data are expressed as the percent of students whose teachers responded in each of the various categories.

Note: The data for grade 12 are from the subset of twelfth-grade students enrolled in a mathematics course at the time the 1992 NAEP mathematics assessment was administered.

Source: National Center for Education Statistics, *Data Compendium for the NAEP 1992 Mathematics Assessment of the Nation and the States*, 1993

questions on the NAEP tests that required writing were difficult for students. Note the difference in the perception of eighth-grade students and their teachers in the amount of time devoted to these two activities. For example, 62 percent of the students report that they never, or hardly ever, write a few sentences about how they solve a problem, while only 41 percent of the students have teachers who responded this way. It is apparent that many students in both grade 8 and 12 still see mathematics as a task from a textbook that they are shown how to do.

Although writing was not part of the reported culture of the mathematics classroom, teachers did say they allowed time for discussing problems and applications to real life. About three-fourths of fourth-grade and eighth-grade students have teachers who report they afford the opportunity to discuss mathematics. It is not clear from the NAEP data exactly what is meant by these "discussions," which could mean anything from simply giving answers to the rich sharing of strategies of solutions. However, it is evident that teachers believe that talking about mathematics is important.

Business and industry repeatedly state the need for employees who can work together to solve problems. There is some evidence from the NAEP data, as reported in table 4.11, that students are involved at least weekly in working in small groups. More than half of the fourth- and eighth-grade students have teachers who report they use group activities in mathematics weekly. Note the difference in the response level of students and teachers to this question. For example, while 38 percent of the eighth-grade students report that they never, or hardly ever, work in small groups, only 17 percent of those students have teachers who report this. Students in all three grades were consistent in their perceptions of how much group work they were doing. One may speculate that students think they still are working alone when placed in a small-group setting. It would be interesting to interview students and teachers to help understand these different points of view on the amount of time spent in group work.

Moreover, business and industry increasingly are requiring their employees to write and defend their findings and positions. Table 4.11, which presents the practices of writing mathematics reports and doing mathematics projects, addresses these skills. Students in grade 4 were not asked to respond to this question, and as with all the NAEP data, no information was gathered from twelfth-grade teachers. However, a consistent pattern in teachers' practices can be discerned from the perceptions of the eighth-grade students. More than three-fourths of students at all three grade levels have never, or hardly ever, participated in writing reports or doing mathematics projects. Usually these types of activities are considered to be long term—not ones to be completed in less than a week—so it is not surprising that they are not frequently done. It is disappointing that most students have had no experience with these activities. As more textbooks

Table 4.11 Teachers' and Students' Reports on Frequency of Small-Group Activities and Projects

	At Least Weekly		Less than Once a Week		Never or Hardly Ever	
	Teachers' Reports[a]	Students' Reports	Teachers' Reports[a]	Students' Reports	Teachers' Reports[a]	Students' Reports
Work in Small Groups						
Grade 4	63	37	28	19	9	44
Grade 8	51	36	32	26	17	38
Grade 12 (taking math)	—	42	—	22	—	36
Write Reports or Do Math Projects						
Grade 4	1	—	17	—	82	—
Grade 8	1	5	21	18	78	77
Grade 12 (taking math)	—	3	—	15	—	82

[a]These data are expressed as the percent of students whose teachers responded in each of the various categories.

Note: The data for grade 12 are from the subset of twelfth-grade students enrolled in a mathematics course at the time the 1992 NAEP mathematics assessment was administered.

Source: National Center for Education Statistics, *Data Compendium for the NAEP 1992 Mathematics Assessment of the Nation and the States*, 1993

integrate such activities into their materials, it will be interesting to note whether they become more common activities for students.

The use of manipulatives was recommended in the first popular textbook for elementary students in the United States (Colburn 1837). Both the student and teacher NAEP questionnaires at grades 4 and 8 gathered information about the frequency of using manipulatives today. The questionnaire asked about the amount of use of some manipulatives as well as measurement instruments such as rulers and geometric models. The responses of both teachers and students are given in table 4.12.

Only 28 percent of the students in grade 8 say they use manipulatives at least weekly, whereas the teachers of 50 percent of the students in grade 8 say they use manipulatives that often. Similar contrasts can be observed between the perception of fourth-grade students and that of fourth-grade teachers. There could be many reasons for the difference in perception. If teachers are using the manipulatives for demonstration purposes but

Table 4.12 Teachers' and Students' Reports on Frequency of Use of Manipulatives

	Almost Every Day		At Least Once a Week		Less than Weekly	
	Teachers' Reports[a]	Students' Reports	Teachers' Reports[a]	Students' Reports	Teachers' Reports[a]	Students' Reports
Grade 4	44	34	46	24	10	41
Grade 8	8	20	50	28	42	52

[a]These data are expressed as the percent of students whose teachers responded in each of the various categories.

Source: National Center for Education Statistics, *Data Compendium for the NAEP 1992 Mathematics Assessment of the Nation and the States,* 1993

students are not using them simultaneously, then students may perceive that they are not actually using manipulatives. The percents may not be that far apart in reality if one considers the difficulty of remembering whether they were used less than once a week or at least weekly. Or, perhaps the difference is caused by the language of *at least weekly* and *less than once a week.* No matter what the cause of the difference between teacher and student perceptions of the frequency of the use of manipulatives, the more important consideration is how the manipulatives are used. Students may need these tools to help develop mathematical concepts and skills or to apply mathematics.

If we expect teachers to use a variety of materials such as manipulatives, calculators, and computers, then we need to be certain that these resources are available. Table 4.13 includes NAEP data on teachers' perceptions on availability of resources. Overall, about half of the students have teachers who report getting most of the teaching resources they need, but very few get all that they think they need. The data in the table also reveal there is not an even distribution of resources according to the location of schools—advantaged urban, disadvantaged urban, and extreme rural. At both grade levels, about half of the students in urban disadvantaged settings have teachers who respond that they lack necessary resources. In the other settings, more than half of the students in grade 4 and a little less than half of the students in grade 8 have teachers who say they get most of the resources they need.

Instructional Assessment

The *Assessment Standards for School Mathematics* (NCTM 1995) calls for many shifts in our view of, and practices with, assessment. There is also much fervor on the part of states and districts as they experiment with

Table 4.13 Teachers' Reports on Availability of Resources

Question: How well supplied are you by your school system with the instructional materials and other resources you need to teach your class?

	I get some or none of the resources I need	I get most of the resources I need	I get all the resources I need
Grade 4			
Nation	36	52	11
Advantaged Urban	32	57	10
Disadvantaged Urban	55	37	8
Extreme Rural	36	50	14
Grade 8			
Nation	33	53	14
Advantaged Urban	29	47	24
Disadvantaged Urban	51	37	12
Extreme Rural	35	46	19

Note: The data in the table are expressed as the percent of students whose teachers responded to each of the various categories.

Source: National Center for Education Statistics, *Data Compendium for the NAEP 1992 Mathematics Assessment of thc Nation and the States,* 1993

different ways to assess student learning for different purposes, such as for monitoring student progress or making instructional decisions. The NAEP questionnaires provide some information about the status of assessment as seen by teachers and students.

Responses of students on the NAEP student questionnaires indicate that there has been a significant change in the amount of testing from 1990 to 1992. Only 30 percent of fourth-grade students reported they took weekly tests in 1992, compared to 43 percent in 1990. Similarly, 55 percent of eighth-grade students and 57 percent of twelfth-grade students who were taking mathematics said they had weekly tests in 1992, compared to 65 percent and 68 percent in 1990. This will be an interesting trend to follow through the next few assessments. As teachers use ways other than tests to assess, the amount of time spent in class on tests alone should change. Even with the change to less time for testing, assuming that tests at grade 12 last a class period, half of twelfth-grade students are spending more than six weeks of class time in testing situations.

Even more disturbing than the amount of time is the differentiation between student groups. For example, in grade 4 more than half of the students in the bottom one-third of the schools are tested daily or weekly, while about only one-fourth of the students in the top one-third of the schools are tested that often. Similar differences are found between other age groups and between demographic subgroups, as shown in chapter 3.

As assessments such as NAEP change on the national and state scene, it is imperative for students to have classroom experiences preparing them not only for the content but also for the format of these new assessments. The data in table 4.14 address several types of assessment activities. For example, the majority of students in grades 4 and 8 had teachers who reported they most often assessed with problem sets, although many students were asked to write short responses. There is some use of multiple-choice tests and presentations, portfolios, and projects. However, more than one-third of the students hardly ever write even short responses on mathematics tests. High school teachers seem to use alternative methods less often than those in the lower grades. According to Weiss (1994), high school teachers report that they feel less prepared to use—and see less need for—alternative assessments than do middle school and elementary teachers.

Students' perceptions of the types of tests that they were experiencing are also revealing. They were asked how often they took tests that included problems that required detailed solutions to mathematics problems that had not been worked on previously. Although a small percentage of students responded that they did so almost daily, only 27 percent, 19 percent, and 11 percent of students in grades 4, 8, and 12, respectively, said they did so at least weekly. Others hardly ever had this opportunity. Overall, the assessments are dominated by what is described in the questionnaire as short written responses and problem sets rather than multiple-choice tests or alternative assessment procedures such as portfolios, projects, or presentations. This suggests that the tests require responses to problems similar to those that students previously have had the opportunity to practice—problems that can be answered quickly and rather routinely. If problem solving, communicating, and reasoning are important in mathematics, these processes must be reflected in classroom assessment.

SUMMARY

The *Professional Standards for Teaching Mathematics* (NCTM 1993) rests on two assumptions: (1) teachers are key figures in changing the ways in which

Table 4.14 Teachers' Reports on Frequency and Type of Testing

	Percent of Students		
	Once or Twice a Week	Once or Twice a Month	Yearly or Never
Multiple-Choice			
Grade 4	6	43	51
Grade 8	4	30	66
Problem Sets			
Grade 4	53	39	9
Grade 8	58	32	10
Short Written Responses			
Grade 4	44	16	40
Grade 8	44	22	33
Projects, Portfolios, or Presentations			
Grade 4	20	25	54
Grade 8	21	32	47

Source: National Center for Education Statistics, *Data Compendium for the NAEP 1992 Mathematics Assessment of the Nation and the States*, 1993

mathematics is taught and learned in schools and (2) such changes require that teachers have long-term support and adequate resources. Who are these teachers so key to changing the ways that mathematics is taught in schools? In this chapter, we have seen a national profile of our teaching force in mathematics at grades 4 and 8 and in the secondary school. Although this profile has been described nationally, we know that it varies from state to state, district to district, school to school, and class to class. Our own knowledge about teachers helps us to put the national data into perspective. No matter what our role in education, it is evident from the national data that there are changes that we need to make. We need to encourage young people of all racial and ethnic groups to teach mathematics. We need to provide in-depth preservice and in-service opportunities for all teachers. We need to provide the conditions that enable teachers to teach in the ways they value. Finally, we need to provide opportunities for all students to learn mathematics in a manner that will prepare them for further schooling and further learning throughout their lives.

The picture of classroom instruction, in many ways, is more sketchy than the profile of teachers' backgrounds. It is difficult to elicit a true picture of an action-packed classroom from a written questionnaire. Quantitative information alone will not point the way to successful practices. These data, we hope, will inspire us to probe deeper into what practices or activities make a difference in the classroom. As we reflect on this NAEP information, our own classroom practice, and that of others, we will continue to learn about practices that will make a difference for our students.

REFERENCES

Bell, Janice P. *An Analysis of the Relationship of Teacher Certification with the Mathematics Achievement of Middle School Students.* (Orangeburg, S.C.: South Carolina State College, 1992.

Blank, Rolf K., and Doreen Gruebel. *State Indicators of Science and Mathematics Education 1995.* Washington, D.C.: Council of Chief State School Officers, 1995.

Colburn, Warren. *Intellectual Arithmetic upon the Inductive Method.* Concord, N.H.: Oliver L. Sanborn, 1837.

Dossey, John A., Ina V. S. Mullis, Steven Gorman, and Andrew S. Latham. *How School Mathematics Functions: Perspectives from the NAEP 1990 and 1992 Assessments.* Washington, D.C.: National Center for Education Statistics, 1994.

Glass, Gene V., Leonard S. Cahen, Mary Lee Smith, and Nicola N. Filby. *School Class Size: Research and Policy.* Beverly Hills, Calif.: Sage Publications, 1982.

Leitzel, James R. C. *A Call for Change: Recommendations for the Mathematical Preparation of Teachers of Mathematics.* Washington, D.C.: Mathematical Association of America, 1991.

National Center for Education Statistics. *Data Compendium for the NAEP 1992 Mathematics Assessment of the Nation and the States.* Report no. 23-ST-04. Washington, D.C.: National Center for Education Statistics, 1993.

National Council of Teachers of Mathematics. *Assessment Standards for School Mathematics.* Reston, Va.: National Council of Teachers of Mathematics, 1995.

_____. *Curriculum and Evaluation Standards for School Mathematics.* Reston, Va.: National Council of Teachers of Mathematics, 1989.

_____. *Professional Standards for Teaching Mathematics.* Reston, Va.: National Council of Teachers of Mathematics, 1991.

Weiss, Iris R. *A Profile of Science and Mathematics Education in the United States: 1993.* Chapel Hill, N.C.: Horizon Research, 1994.

5

What Do Students Know about Numbers and Operations?

Vicky L. Kouba, Judith S. Zawojewski, & Marilyn E. Strutchens

THE CONTENT area of numbers and operations remains a substantial part of most school mathematics curricula and for the general public constitutes what is meant by the term *mathematics*. A large portion of the mathematics items for the 1992 NAEP assessment fall within the topic of numbers and operations: that is, 40 percent of the mathematics items given to fourth-grade students, 30 percent of those given to eighth-grade students, and 25 percent of those given to twelfth-grade students. This chapter provides information on what our students know and can do regarding concepts and operations for both whole numbers and rational numbers.

WHOLE NUMBERS

The mathematics strand of whole numbers continues to dominate the mathematics curriculum and, as such, is an area where students do well on tasks set in familiar or routine contexts. The next sections present results on NAEP items concerning whole-number concepts and properties including place value, rounding, representation and comparison of whole numbers, number theory, and whole-number operations and applications in both numerical and word problem contexts.

Whole-Number Concepts and Properties

The 1992 NAEP contained twenty-one items, each administered at either one grade level or across grade levels, that assessed performance on place value, rounding, representations of numbers, and number theory. In general, students appear to have an understanding of place value, rounding, and number theory that serves them well in familiar, straightforward contexts. However, their understanding may not be sufficiently well

HIGHLIGHTS—WHOLE NUMBERS

- Students at all three grade levels appear to have an understanding of place value, rounding, and number theory concepts in familiar, straightforward contexts but have difficulty applying the concepts and properties to unfamiliar or complex situations.
- Whole-number addition calculations give little trouble to students at all three grade levels.
- Most students at all three grade levels performed well on subtraction calculations involving regrouping with no zeros as digits.
- Most eighth-grade students did well on one-step and routine two-step word problems involving addition and subtraction.
- Fourth-grade students did better on one-step multiplication word problems than on one-step division word problems. Eighth- and twelfth-grade students did well on both one-step multiplication and division word problems.
- Performance for eighth- and twelfth-grade students on two-step and multistep multiplication and division items was mixed and seemed to depend on the item's complexity or its context.

developed to help students apply the concepts and properties to more unfamiliar or complex situations. Performance on items in these topics in whole numbers is discussed in the sections that follow.

Place Value

Eight items directly addressed the understanding of place-value concepts, and performance data for those items appear in table 5.1. All eight items required students to demonstrate an understanding of place value beyond merely identifying which digit was in a particular place. In particular, two items were in extended constructed-response format, which required students to demonstrate their reasoning and problem-solving abilities.

At all three grade levels, students can identify numbers on the basis of stated place values or changes in place value, for example, knowing the difference between the largest three-digit number and the smallest three-digit number. Performance on those kinds of place-value items was, for the most part, at or above the 50-percent-correct level, as shown for items 1, 2, and 5 in the table. However, it was somewhat surprising that students in grade 12 had some difficulty working with changes in more than one place value in multidigit numbers, as shown by the percent-correct value for item 8 in the table. Eighth-grade students had difficulty working with large numbers, as illustrated by the percent-correct value for item 6. Both fourth- and

Table 5.1 Performance on Place-Value Items

Item Description	Percent Correct		
	Grade 4	Grade 8	Grade 12
1. Choose a number that is 10 more than another number.	68	—	—
2. Write a number, given the place value for each digit.	54	60	—
3. Choose the number representing the change in 217 if the digit 1 were replaced by a digit 5.	36	72	—
4. Demonstrate an understanding of place value; explain.	20[a]	—	—
5. Choose the number representing the difference between the smallest positive 3-digit and the largest positive 2-digit numbers.	—	58	68
6. Choose the number of millions in one billion.	—	22	—
7. Reason about how to maximize the difference in a subtraction problem based on place value.	—	13[a]	—
8. Choose correct description of a number resulting from changes in place value.	—	—	49

[a]Percent of students scoring at either the satisfactory or extended level

Note: Item 2 was a regular constructed-response item, items 4 and 7 were extended constructed-response items, and the rest were multiple-choice items.

eighth-grade students had low performance levels on place-value items presented in extended constructed-response format. In particular, those students could not justify their answers or explain their reasoning.

A closer examination of responses to some of the place-value items provides some insight into the particulars of student performance. Item 1 in table 5.2 illustrates the success that fourth- and eighth-grade students had working with changes in place value, here in the tens place. Slightly more than a third of the fourth-grade students and almost three-fourths of the eighth-grade students responded correctly with an answer of 40. The encouraging aspect of performance on that item is that although 14 percent of the fourth-grade students selected 400, a number that is unreasonably

Table 5.2 Whole-Number Place Value: Sample Items

Item	Percent Responding		
	Grade 4	Grade 8	Grade 12
1. By how much would 217 be increased if the digit 1 were replaced by a digit 5?			
A. 4	22	12	—
B. 40*	36	72	—
C. 44	21	11	—
D. 400	14	3	—
2. What is the difference between the smallest positive 3-digit number and the largest positive 2-digit number?			
A. 1*	—	58	68
B. 9	—	6	3
C. 10	—	13	11
D. 90	—	8	6
E. 900	—	12	9
3. A certain reference file contains approximately one billion facts. About how many million is that?			
A. 1,000,000	—	40	—
B. 100,000	—	10	—
C. 10,000	—	5	—
D. 1,000*	—	22	—
E. 100	—	23	—

*Indicates correct response.

Note: Percents may not add to 100 because of rounding or omissions.

large in the context of the item, only 3 percent of the eighth-grade students did so. This decrease may be interpreted as a growth in number sense about place value from grade 4 to grade 8. However, students' understanding of place value in relatively simple contexts may not carry over into more complex situations. For example, on a secure multiple-choice item that was similar to item 1 in table 5.2 but involved finding how much a number changed when *two* digits were replaced, only 49 percent of the twelfth-grade students responded correctly, while 7 percent chose a number representing a change too small to be a reasonable answer. Sowder suggests that students have difficulty with many aspects of place-value relationships because much of the present curriculum "develops little meaning beyond what is required to read and write numbers, identify the place value of the digits that appear in numbers (e.g., there are 5 tens in 356) and expand numbers (e.g., 356 is 3 hundreds + 5 tens + 6 ones)" (1992, p. 10).

The two place-value items, labeled item 2 and item 3 in table 5.2, were given to eighth-grade students, with one also given to twelfth-grade students. Results for item 2, which required eighth- and twelfth-grade students to find the difference between the smallest positive three-digit number and the largest positive two-digit number, show that almost 60 percent of the eighth-grade students and almost 70 percent of the twelfth-grade students responded correctly. The results for item 3, which required eighth-grade students to identify how many millions are in a billion, suggest that students continue to have difficulty representing and thinking about large numbers, a finding consistent with recent research on students' number sense (Sowder and Kelin 1993). Only 22 percent of the eighth-grade students identified 1,000 as the correct response, and 40 percent of the students chose the distracter 1,000,000. Although it could be that some students ignored the information about the "one billion facts" in the first part of item 3 and merely translated the word *million* into its numerical representation, it is also possible that others thought there were a million millions in a billion. This implies that students have little sense of the relative size of large numbers and how those numbers are represented in the base-ten number system. These students also may lack an understanding of the multiplicative nature of the place-value system.

Rounding

Four items assessed students' performance on rounding whole numbers, with one item administered only to fourth-grade students, one item only to eighth-grade students, and the remaining two items administered to both fourth- and eighth-grade students. The items all involved the traditional, school-taught strategies of rounding to the nearest given place value such as tens, hundreds, thousands, and so on. In general, students were successful on rounding items presented in traditional ways, with percent-correct values ranging from 60 percent for fourth-grade students to 83 percent for eighth-grade students. Performance was lower on items presented in less traditional ways or in nonroutine formats, with percent-correct values ranging from 22 percent to 44 percent for fourth-grade students and from 49 percent to 69 percent for eighth-grade students.

Item 1 in table 5.3 illustrates a routinely worded rounding task, and item 2 illustrates a nonroutine rounding task. Although 60 percent of the fourth-grade students chose the correct response for the routine item, one-fourth chose the incorrect response of rounding to the nearest hundred instead of to the nearest thousand. While some students may have just guessed, others may have either failed to read the problem carefully or, more conceptually problematic, failed to correctly distinguish the thousands place from the hundreds place. For the nonroutine item, fewer than one-fourth of the fourth-grade students and about half the eighth-grade students wrote a correct response. That 26 percent of the fourth-grade students and 17 per-

Table 5.3 Rounding Whole Numbers: Sample Items

Item	Percent Responding	
	Grade 4	Grade 8
1. What is 18,565 rounded to the nearest thousand?		
A. 18,000	8	—
B. 18,600	24	—
C. 19,000*	60	—
D. 20,000	6	—
2. The length of a dinosaur was reported to have been 80 feet (rounded to the nearest 10 feet). What length other than 80 feet could have been the actual length of this dinosaur?		
Correct response in the range $75 \le n \le 85$	22	49
Incorrect response of 90	26	17
Any incorrect response other than 90	48	30
Omitted	4	4

*Indicates correct response.
Note: Percents may not add to 100 because of rounding or omissions.

cent of the eighth-grade students gave "90" as an answer may indicate that some students misread or misinterpreted the item as one that restricted answers to those expressed as multiples of 10. In this instance, it would have been helpful to know how many students responded "70 feet"; however, the NAEP scoring guide did not allow for coding this numerical answer. Performance on this nonroutine item suggests that, while researchers may need to explore the roles and interactions of linguistic knowledge (reading comprehension) and conceptual knowledge (the meaning of rounding), teachers may need to explore the usefulness of recommended teaching actions that help students make sense of nonroutine problems (Kroll and Miller 1993). Because item 2 involves several aspects of number sense, rounding, and conceptual and linguistic knowledge, it may provide a rich context for discussion in middle school classrooms.

Performance on a secure rounding item presented in a real-world context and involving the difference between two weight measurements demonstrated that more than one-fourth of students in grade 4 had difficulty identifying the correct operation to use. Rather than choosing the result of rounding the *difference* of two numbers, 29 percent of the fourth-grade students chose the result of rounding the *sum* of the two numbers. This is consistent with a pattern appearing throughout the 1992 NAEP data that many fourth-grade students often incorrectly employ a "when in doubt, add" strategy (see chapter 6). Kroll and Miller (1993) classify this strategy as an aspect of poorly developed conceptual knowledge. On an encouraging note, fewer than 10 percent of students in grade 8 chose the distracters associated with the incorrect arithmetic operation.

Representing and Comparing Whole Numbers

Only two items on the 1992 NAEP involved representing and comparing whole numbers using place value. One item asked fourth- and eighth-grade students to select from a list of choices the correct numerical representation for "three hundred fifty-six thousand, ninety-seven" (356,097). Overall, students did well on that item, with 72 percent of the fourth-grade students and 89 percent of the eighth-grade students responding correctly. This is especially encouraging given that the item required recognizing that zero must serve as a place holder when expressing the number numerically.

The other item was a secure item asking fourth-grade students to identify which of four multidigit numbers was the greatest. The answer choices for that item involved comparisons with zero as a place holder. Seventy-three percent of the fourth-grade students who were administered the item responded correctly. These results are comparable to those on a similar multiple-choice item from the 1990 NAEP that required students to choose which of the following numbers was greatest: 2,573; 2,537; 2,753; or 2,735.

In 1990, 81 percent of the fourth-grade students chose the correct answer. Although these performance results are encouraging, a discouraging aspect is that on both items about 20 percent to 25 percent of the students either guessed or chose an incorrect answer on the basis of some misconceptions about place value and comparing numbers. These results suggest that by grade 4 some students do not understand place value, especially the role of zero, when comparing numbers.

Number Theory

In the NAEP framework, number theory concepts included on the assessment included odd and even numbers at all three grades and the following concepts at grades 8 and 12 only: multiples including least common multiple (LCM) and divisors including greatest common divisor (GCD), prime numbers, factorization, divisibility, and remainders (National Assessment of Educational Progress 1988). The 1992 NAEP included eight items related to number theory concepts, and these items were administered only to students in grades 8 and 12. Table 5.4 contains a summary of performance on those eight items.

Table 5.4 Performance on Number Theory Items

| | Percent Correct | |
Item Description	Grade 8	Grade 12
1. Choose a number that is both a multiple of 3 and 7.	77	88
2. Interpret a rule about odd and even numbers.	39	51
3. Choose an operation that always results in odd integers.	27	—
4. Find the first three terms in a given sequence.	—	40
5. Evaluate an expression for odd or even numbers.	—	38
6. Choose the sum of the least values that makes a given equality true.	—	30
7. Apply concepts of multiples and choose the correct remainder.	—	24
8. Explain and generalize a given number pattern based on squaring numbers that end in the digit 5.	—	2[a]

[a]Percent of students scoring at either the satisfactory or extended level

Note: Items 4 and 5 were regular constructed-response items, item 8 was an extended constructed-response item, and the rest were multiple-choice items.

In general, eighth- and twelfth-grade students had little trouble with items that assessed number theory concepts in a straightforward way, such as asking students to choose from a list of numerical values the one that is a multiple of two given numbers (item 1 in table 5.4). However, most number theory items on the 1992 assessment were complex problems set in abstract, mathematical contexts, and performance on those items was very low, ranging from 27 percent to 39 percent correct at grade 8 and from 24 percent to 51 percent correct at grade 12. Item 8, an extended constructed-response question that required students to construct a proof about squares of numbers, was extremely difficult for students in grade 12.

The items in table 5.5 illustrate the extremes in performance on number theory items. Students in grades 8 and 12 had little difficulty in choosing the correct answer for item 1, a simple, straightforward question. That students were permitted to use a calculator may have contributed to the high performance levels. Although students also had a calculator for item 2, a factorization problem set in an abstract mathematical context, only about one-third of twelfth-grade students selected the correct answer. Twenty-three percent of the students selected 17, the number obtained by adding the given numbers and ignoring the variables. Although this result might be attributed to guessing, it might also suggest that for complex items, some students in grade 12 may have applied a naive strategy of "when in doubt, add."

Whole-Number Operations and Applications

The 1992 NAEP mathematics assessment contained 39 items that directly assessed students' performance on addition, subtraction, multiplication, and division of whole numbers in both number contexts and word (or applied) problems. The recognition of appropriate multiplication and division number sentences was also assessed. Performance varied across grade level and problem type. Fourth-grade students' performance ranged from 45 percent to 89 percent correct on computational items presented in number contexts. They did best on addition items and routine items involving other operations and for which calculators were available. Performance for students in grade 4 fell on nonroutine items and items requiring a fully developed concept of the nature and properties involving zero, and on word problems, with percent-correct values ranging from 31 percent to 67 percent.

About 80 percent to 90 percent of the eighth-grade students and slightly more than 90 percent of the twelfth-grade students have apparently mastered straightforward computation and can solve word problems, including routine two-step problems. Performance dropped for eighth- and twelfth-grade students on nonroutine problems, with performance rang-

Table 5.5   Sample Number Theory Items

	Percent Responding	
Item[a]	Grade 8	Grade 12
1. Which of the following is both a multiple of 3 and a multiple of 7?		
A. 7,007	7	4
B. 8,192	4	2
C. 21,567*	77	88
D. 22,287	5	3
E. 40,040	3	1
2. Suppose $4r = 3s = 10t$, where r, s, and t are positive integers. What is the sum of the least values for r, s, and t for which this equality is true?		
A. 7	—	10
B. 17	—	23
C. 41*	—	30
D. 82	—	9
E. 120	—	15

*Indicates correct response.

[a]Items are from item blocks for which students were provided with, and permitted to use, scientific calculators.

Note: Percents may not add to 100 because of rounding or omissions.

ing from below 15 percent correct on complex problems requiring the application of fully developed number sense and the ability to explain answers to between 70 percent and 80 percent correct on nonroutine problems requiring two or more computational steps and some comparison of numbers.

Addition and Subtraction

Table 5.6 contains performance data for the set of addition and subtraction items. Four items, one involving addition and three involving subtraction, assessed those operations in numerical contexts, and each item involved regrouping. Results on the addition item, described as item 1 in the table, showed that whole-number addition gives students at all three grade levels little trouble. Eighty-eight percent of the fourth-grade students and 93 percent of the eighth- and twelfth-grade students answered the problem correctly. The most common error appeared to be choosing a response derived from an incorrect regrouping process. Performance on the subtraction items, described as items 2, 3, and 4 in the table, varied and was a little lower than on the addition item. Fourth-grade students appeared to have difficulty with the presence of zeros as digits. It is encouraging, however, to see that by grade 8 those particular difficulties with zero have been overcome, as 84 percent of the eighth-grade students chose the correct solution.

Table 5.6 Performance on Whole-Number Addition and Subtraction Items

Item Description	Percent Correct		
	Grade 4	Grade 8	Grade 12
Numerical Contexts			
1. Add two three-digit numbers with regrouping and choose the correct sum.	88	93	93
2. Subtract a one-digit number from a two-digit number.	76	—	—
3. Subtract a two-digit number from a three-digit number.	73	86	—
4. Choose the correct solution to 503 – 207.	53	84	—
Applied Contexts			
5. Choose the correct answer to a one-step subtraction problem involving two-digit numbers.	67	92	—
6. Solve a one-step subtraction problem involving three-digit numbers.	55	—	—
7. Estimate which two food items would provide a total of about 600 calories.	45	—	—
8. Solve a multistep problem involving two three-digit numbers.	33	75	—
9. Solve a nonroutine, multistep problem and explain.	—	31	50

Note: Items 1, 4, and 5 were multiple-choice items, and the rest were regular constructed-response items.

Five NAEP items addressed addition and subtraction in applied contexts and are described in table 5.6 as items 5 through 9. All items except for the estimation item (item 7) involved regrouping. Response patterns indicate that about half to nearly a third of the fourth-grade students can solve one-step word problems involving addition and subtraction. Performance dropped to about 33 percent for fourth-grade students on a relatively routine two-step problem, described as item 8 in the table, perhaps because of the need to recognize and correctly perform two calculations. Eighth-grade students did well on routine one- and two-step items, with percent-correct values of 75 percent and 92 percent. However, performance for eighth- and twelfth-grade students was 31 percent and 50 percent, respectively, on item 9, a nonroutine item that involved determining the relative change in the size of addends when a sum is held constant. This item may have been difficult for students because it was more conceptually complex than a straightforward computation item; but, because the item was a regular constructed-response item and students' incorrect responses were not analyzed as part of NAEP reporting, the actual difficulties cannot be identified.

An examination of the two released applied addition and subtraction items offers further information on students' understanding. Item 1 in table 5.7 was the only one of the five items for which students were provided with, and permitted to use, calculators. The item required fourth-grade students to identify which two of four menu items would have a sum of *about* 600 calories. Forty-six percent of the fourth-grade students gave acceptable answers to the item, with 44 percent identifying the cheeseburger and yogurt as having a combined sum of about 600 calories and another 2 percent of the students indicating that two hot dogs would also satisfy the calorie constraint. Data linking the response to the self-report question on calculator use ("Did you use the calculator on this question?") to performance on item 1 reveals that students who reported using the calculator had a significantly higher percent-correct value, 58 percent, than students who reported not using the calculator, 36 percent. This may be the result of higher-performing students' electing to use a calculator and lower-performing students' electing not to use a calculator, or it may suggest that using calculators helps increase students' accuracy in computations with multidigit numbers. The pattern of higher performance by groups of students reportedly using calculators on multidigit items was repeated for other NAEP items (see chapter 8). Because NAEP provided no information about the kinds of incorrect responses fourth-grade students gave or about how students used the calculator, it might be interesting to give this item to students and then analyze their responses for error patterns, ways in which students used or did not use the calculator, or important misconceptions such as failing to take into account all conditions in the item before answering.

Students did considerably better on item 2 in table 5.7 than on item 1, with 67 percent of the fourth-grade students and 92 percent of the eighth-grade students choosing the correct response. This item may have been easier for several reasons. First, it is a multiple-choice item rather than a constructed-response item, so some students may have guessed the correct answer. The item is also a bit more straightforward because it involves only two numbers and one operation rather than four numbers, an operation, and an estimation or rounding procedure required in item 1. How-

Table 5.7   Whole-Number Addition and Subtraction Word Problems

Item	Percent Responding	
	Grade 4	Grade 8

Cheeseburger	**Hot Dog**	**Yogurt**	**Cookie**
393 Calories	**298 Calories**	**214 Calories**	**119 Calories**

Item	Grade 4	Grade 8
1. Which two of the items above would provide a total of about 600 calories?		
Correct response of cheeseburger, yogurt	44	—
Correct response of 2 hot dogs	2	—
Any incorrect response	49	—
Omitted	5	—
2. There are 50 hamburgers to serve 38 children. If each child is to have at least one hamburger, at most how many of the children can have more than one?		
A. 6	11	3
B. 12*	67	92
C. 26	14	3
D. 38	6	1

*Indicates correct response.

Note: Percents may not add to 100 because of rounding or omissions.

ever, the level of performance is encouraging because the item does not contain key words such as *difference* that could cue the students into using a subtraction operation. It may be that some students saw this item not as a subtraction problem but instead as a division problem, with the remainder of 12 being the correct answer. Because of the item's multiple-choice format, no information is available from NAEP on whether students chose to subtract or divide or whether the choice of operation differed between grade levels. Giving this item with or without the answer choices to elementary or middle school students could provide additional information on students' solution strategies.

Multiplication and Division

Four NAEP items involved using number sentences to represent multiplicative situations, six items addressed the use of multiplication and division in numerical contexts, and eighteen items set in applied contexts as word problems required the use of multiplication or division and may have also required the use of addition and subtraction. Performance varied across items depending on the size of the numbers involved, the availability of calculators for use in solving, and the complexity of the item.

The four items requiring that students choose the correct number sentence to represent each problem were given to fourth-grade students only. Seventy-four percent of the fourth-grade students chose a correct multiplication sentence from among addition, subtraction, multiplication, and division sentences for a situation involving a straightforward basic number fact. Performance fell to about 45 percent on items that included extraneous numbers or that required two steps. Only 37 percent of the fourth-grade students correctly solved item 1 in table 5.8. The numbers used in the item may have misled students into choosing a missing factor (division) sentence rather than the correct multiplication sentence. That is, because the numbers 6 and 24 are associated with the commonly practiced number fact $6 \times 4 = 24$, some students—and in the case of NAEP, 27 percent—may have interpreted the item as $6 \times \square = 24$ rather than $6 \times 24 = \square$. Such a misinterpretation would not be unusual, according to Graeber and Tanenhaus (1993), who found that students often "key" on the numbers in a problem rather than rely on the meaning of the problem when solving it. One must be cautious, however, in making any firm conclusions on the basis of item 1 because 24 percent of the students omitted it.

Five of the six multiplication and division items presented in a numerical context were given to fourth-grade students only, while the sixth item was given to eighth-grade students only. There were no numerical-context multiplication and division items on the grade-12 NAEP assessment in 1992. Eighty-nine percent of the fourth-grade students correctly solved the regular constructed-response item "$3 \times 405 =$ " and another such item, "Divide

Table 5.8 Whole-Number Multiplication and Division Items: Number Sentences and Number Contexts

Item	Percent Responding Grade 4
1. Marlene made 6 batches of muffins. There were 24 muffins in each batch. Which of the following number sentences could be used to find the number of muffins she made?	
A. $6 \times \square = 24$	27
B. $6 + 24 = \square$	9
C. $6 + \square = 24$	3
D. $6 \times 24 = \square$ *	37
2. In the multiplication problem below, write the missing number in the box.	
$$\begin{array}{r} 23\square \\ \times\ \ 8 \\ \hline 1{,}896 \end{array}$$	
Correct response of 7	58
Incorrect response of 2	11
Incorrect response of 8	4
Any other incorrect response	18
Omitted	9

*Indicates correct response.
Note: Percents may not add to 100 because of rounding or omissions.

108 by 9." Eighty-two percent of the fourth-grade students also responded correctly to a secure multistep, multiple-choice item involving multiplication and division. Eighty-four percent of the eighth-grade students correctly solved an item involving the division of a three-digit by a two-digit number. Thus, it appears that simple whole-number multiplication and division is not problematic for fourth- or eighth-grade students.

Performance dropped for fourth-grade students when the item involved multiplication by 0 or when the problem was set in an unfamiliar context. Of fourth-grade students responding to an item involving multiplication by 0, 46 percent selected the correct answer, which suggests that those students knew that multiplying any number by 0 results in 0. For the non-routine item labeled item 2 in table 5.8, about 60 percent of the fourth-grade students supplied the correct missing digit. Although this performance is not particularly low, it is lower than on the simpler, numerical-context items described above. Also, because students had a calculator while working on item 2, performance data are available for the self-report question, "Did you use the calculator on this question?" For students who reported using the calculator, the percent-correct value was 82 percent, and for those who reported not using the calculator, the corresponding value was 35 percent, with the difference between the values being statistically significant. However, NAEP results provide no information on how students actually used or did not use the calculator as a tool in item 2. These differences in performance make the item potentially useful for educators interested in comparing strategies between groups who have a calculator available and groups who do not have a calculator available. The item might also prove useful for examining whether students recognize the inverse nature of multiplication and division.

Nineteen NAEP items addressed the application of multiplication and division in word problems. Table 5.9 contains summaries of, and percent-correct values for, a selected set of eleven of the nineteen items to show the range of performance within and among grade levels. Results for fourth-grade students show that they did best on problems involving multiplication rather than division and on simple items that required only one step or calculation rather than more-complex items that involved two or more calculations. Performance levels for students in grade 4 ranged from 55 to 62 percent correct on those one-step items but dropped to a range of 31 to 48 percent correct on word problems that involved unfamiliar contexts, division with remainders, or more than one step or calculation. In general, both eighth- and twelfth-grade students did best on one-step multiplication items or simple division-with-remainders items, with performance ranging from about 50 to 80 percent correct for eighth-grade students and about 70 to 90 percent correct for twelfth-grade students. The most difficult items for students in those grades were extended constructed-response questions and multiple-step items set in nonroutine contexts. An examination of some individual items provides further insight into these general conclusions.

Results associated with the multiple-choice items shown in table 5.10 show that fourth-grade students did better on one-step multiplication items than on one-step division items: 55 percent correct for item 1 as compared to 47 percent and 37 percent correct for items 2 and 3, respectively. The

Table 5.9 Performance on Selected Multiplication and Division Word Problems

Item Description	Percent Correct		
	Grade 4	Grade 8	Grade 12
1. Determine the correct answer to a multistep multiplication problem involving money (cents).	62	82	93
2. Choose the correct answer to a ratio problem involving one-digit numbers.	58	—	—
3. Determine the number of tapes sold in a given number of weeks based on the number of tapes sold per week.	55	—	—
4. Identify the correct procedure to solve a multiplication problem involving two-digit numbers.	48	81	91
5. Determine the correct answer to a problem involving division with remainders.	38	80	92
6. Choose the correct answer to a problem involving division with remainders.	37	69	—
7. Solve a problem involving division with remainders.	31	—	—
8. Solve a multistep problem involving division with remainders.	—	52	72
9. Choose the correct answer to a multistep problem involving more than one operation.	—	43	61
10. Select the correct answer to a multistep, rate vs. time problem involving more than one operation.	—	—	48
11. Determine the correct answer for a nonroutine problem involving exponential growth.	—	—	31

Note: Items 7 and 8 were regular constructed-response items, and the rest were multiple-choice items.

division items may be more difficult because they have the added complexity of requiring students to make a decision about what to do with the remainder from the division operation before determining an answer to the item. The percent of fourth-grade students who responded incorrectly

Table 5.10 Whole-Number Multiplication and Division: One-Step Word Problems

Item	Percent Responding	
	Grade 4	Grade 8
1. A store sells 168 tapes each week. How many tapes does it sell in 24 weeks?		
A. 7	14	—
B. 192	26	—
C. 4,032*	55	—
D. 4,172	3	—
2. Mario sells 80 pencils a day from his supply of 1,000 pencils. Which of the following is true?		
A. He will run out of pencils in 7 days.	16	—
B. He will have enough pencils to sell 80 each day for at least 10 days.*	47	—
C. He will have enough pencils to sell 80 each day for at least 80 days.	16	—
D. He will have enough pencils to sell 80 each day for at least 100 days.	18	—
3. Christy has 88 photographs to put in her album. If 9 photographs will fit on each page, how many pages will she need?		
A. 8	8	3
B. 9	24	21
C. 10*	37	69
D. 11	28	8

*Indicates correct response.

Note: Percents may not add to 100 because of rounding or omissions.

to item 2 in the table were about equally distributed across the three incorrect choices (A, C, and D). For item 3 in the table, choice B (9) and choice D (11) were the most popular incorrect responses. Although the results may be attributed to guessing, they can provide some insight into students' misunderstandings and error patterns. Because 88 divided by 9 equals 9 with a remainder of 7, students may have just dropped the remainder and picked 9 as the answer. As opposed to just guessing, students may have picked 11 pages as an answer because the 88 pictures can be equally distributed across that number of pages. Eighth-grade students also had some difficulty with this item. Twenty-one percent of the eighth-grade students selected choice B (9) as the appropriate response, which suggests that the majority of the eighth-grade students were able to complete the division but were unable to make the correct choice about what to do with the remainder.

The results for item 3 in table 5.10 and the difficulties associated with decisions about remainders are similar to those associated with a 1983 NAEP item that asked students to find the number of buses required to transport 1,128 soldiers if each bus held 36 soldiers (NAEP 1983). In NAEP, only 24 percent of a national sample of 13-year-olds were able to give the correct response. Silver, Shapiro, and Deutsch (1993) concluded that the findings of investigations into students' observed difficulties with division problems involving remainders can be attributed, in part, to "the failure to relate computational results to the situation described in the problem" (p. 118). However, the encouraging news is that because of the 69-percent-correct value for eighth-grade students on the 1992 division-with-remainders item versus the 24-percent-correct value on the 1983 item, there is some evidence of improvement on this aspect of division.

The items shown in table 5.11, together with previously discussed items, show the variability of performance by students, particularly those in grades 8 and 12, on multistep multiplication and division word problems. Performance on item 1 in the table was the highest level achieved by students at all three grade levels on the NAEP multiplication and division word problems: 62 percent for fourth-grade students, 82 percent for eighth-grade students, and 93 percent for twelfth-grade students. These results suggest that in straightforward contexts, even students in grade 4 can successfully handle multistep problems. However, it is unclear whether students perceived the item as an addition problem with seven addends or as a multistep item. For that reason, the item may be an interesting one for educators to use to assess whether students perceive the situation as one that can be done using only addition or a combination of multiplication and addition.

Item 2 in table 5.11 was a multistep item that required students to relate the computation back to the context of the item as a necessary part of finding a correct response, which makes this item similar to the division-with-remainders item discussed previously. Some students may have approached this as a division-with-remainders item by dividing the product of 28 and 64 by 500 to determine the number of 500-sheet reams needed to print 28 copies of a 64-page paper. The most common incorrect responses for both eighth- and twelfth-grade students were "3, or 3.5, 3.6 or about 4." This is consistent with the error pattern of giving a non-whole-number response

Table 5.11 Whole-Number Multiplication and Division: Multistep Word Problems

Item Description	Percent Responding		
	Grade 4	Grade 8	Grade 12

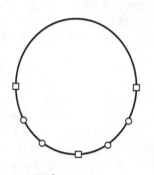

Each □ costs 6¢
Each ○ costs 4¢

1. If the string does not cost anything, how much does the necklace above cost?

A. 10¢	24	10	2
B. 24¢	8	5	2
C. 28¢	5	3	2
D. 34¢*	62	82	93

2[a]. Raymond must buy enough paper to print 28 copies of a report that contains 64 sheets of paper. Paper is only available in packages of 500 sheets. How many whole packages of paper will he need to buy to do the printing?

Correct response of 4	—	52	72
Incorrect response of 3	—	8	7
Incorrect response of 3.5 to 3.6 or "about 4"	—	7	7
Any other incorrect response	—	30	12
Omitted	—	3	2

| | Percent Responding | | |
Item Description	Grade 4	Grade 8	Grade 12
3. It takes 28 minutes for a certain bacteria population to double. If there are 5,241,763 bacteria in this population at 1:00 p.m., which of the following is closest to the number of bacteria in millions at 2:30 p.m. on the same day?			
A. 80	—	—	7
B. 40*	—	—	31
C. 20	—	—	19
D. 15	—	—	32
E. 10	—	—	8

*Indicates correct response.
[a]Item is from an item block for which students were provided with, and permitted to use, scientific calculators.
Note: Percents may not add to 100 because of rounding or omissions.

reported for both the 1983 NAEP assessment and the California Assessment Program (Silver, Shapiro, and Deutsch 1993). However, regardless of the method used, the results are encouraging in that they show both the progress of achievement from grade 8 to grade 12 (52 percent to 72 percent correct) and the progress over the 24 percent correct for 13-year-olds on the NAEP division-with-remainders Bus problem.

Performance on item 3, which was given only to twelfth-grade students, was considerably lower than on item 2, with only 31 percent of those students answering correctly. This item required that students use approximations and identify that there are three doublings of the bacteria because there are 3 groups of 28 minutes in an hour and a half and that they realize that three doublings results in a sixfold increase. The 32 percent of the students who chose D (15) as an answer is greater than would be expected if students were just guessing at answers. While some students may have guessed, others may have correctly determined that three doublings occurred but did not understand or pay attention to the "doubling" condition. This suggests that some students may have had difficulty attending to all the conditions of the complex situation presented in this item.

Problem Solving with Whole Numbers

One of the more complex questions that deals with whole-number operations in an applied context was Treena's Budget, an extended constructed-response question given to eighth-grade students and shown in table 5.12. Solving this problem correctly involved using basic whole-number skills to make decisions about which alternatives Treena could select, based on the money available from her scholarship to basketball camp. While working on this question, students were provided with, and permitted to use,

Table 5.12 Treena's Budget

Question[a]	Percent Responding Grade 8
[General directions]	
This question requires you to show your work and explain your reasoning. You may use drawings, words, and numbers in your explanation. Your answer should be clear enough so that another person could read it and understand your thinking. It is important that you show <u>all</u> your work.	
Treena won a 7-day scholarship worth $1,000 to the Pro Shot Basketball Camp. Round-trip travel expenses to the camp are $335 by air or $125 by train. At the camp she must choose between a week of individual instruction at $60 per day or a week of group instruction at $40 per day. Treena's food and other expenses are fixed at $45 per day. If she does not plan to spend any money other than the scholarship, what are <u>all</u> choices of travel and instruction plans she could afford to make?	
Explain your reasoning.	
Extended response	2
Satisfactory response	2
Partial response	15
Minimal response	22
Incorrect	37
Omitted	22

[a]Question is from an item block for which students were provided with, and permitted to use, scientific calculators.

scientific calculators. Responses to Treena's Budget were scored according to a five-level scheme—incorrect, minimal, partial, satisfactory, and extended—and the designation "no response" (or omitted) for blank papers. Results for this question also appear in table 5.12, with a description of, and an illustrative example for, each performance category appearing in figure 5.1 The descriptions and examples are as they appeared in a NAEP publication (Dossey, Mullis, and Jones 1993, pp. 117–20).

Students in grade 8 did not do well on this extended question, despite the fact that only understanding of whole-number concepts was needed to solve the problem and despite having access to a calculator. Fewer than 5 percent of the students gave extended responses that either completely identified all three options that Treena had with supporting work for each option or gave satisfactory responses that identified the three options (but with incomplete work) or contained correct mathematical evidence for any two of her three options with appropriate supporting work. Another 37 percent wrote partial or minimal responses that were characterized by

---

Incorrect—The work is completely incorrect or irrelevant, or the response states, "I don't know."

> Add everything other
> then scholarship and
> you will get 230.

Minimal—a) Student indicates one or more options only (such as group and train) with no supporting evidence, or b) Student work contains major mathematical errors and/or flaws in reasoning (for example, the student does not consider Treena's fixed expenses).

> She could take the train
> to camp have individual
> instruction and eat everyday
> and not run out of money.

Continued

Partial—The student a) indicates one or more correct options; additional supporting work beyond the minimal level must be present, but the work may contain some computational errors; or b) demonstrates correct mathematics for one or two options, but does not indicate the options that are supported by his or her mathematics.

train at $ 125

```
      280
  40  315      group
  ×7  ───       at
 ─── 720
 280           345
               ×7
               ───
               315
```

She just took the cheapest ones of her choices

now she has money left over

$720 would she all spend

Satisfactory—The student a) shows correct mathematical evidence that Treena has three options, but the supporting work is incomplete; or b) shows correct mathematical evidence for any two of Treena's three options and the supporting work is clear and complete.

125 + 420 + 315 = $ 860

$1000 > $860

$1000 is more than $800 she has money left over so she could take private lessons, a train and her food.

335 + 315 + 280 = 930

$1000 > $930

She could take a plane, her food, and group lessons

Extended—The correct solution indicates what the three possible options are and includes supporting work for each option.

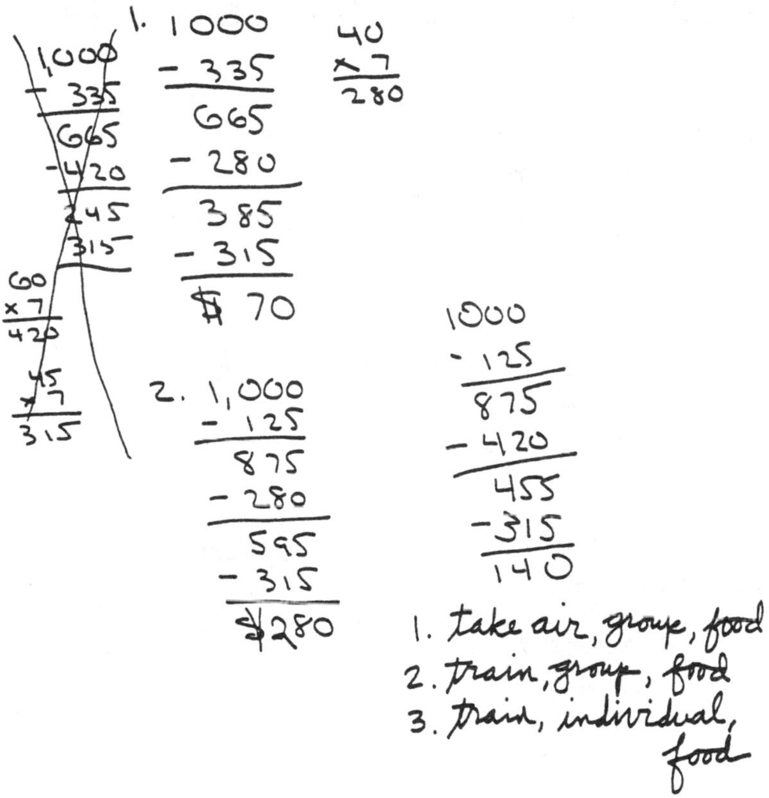

Fig. 5.1 Treena's Budget: Performance categories and sample responses
Source: Dossey, Mullis, and Jones. *Can Students Do Mathematical Problem Solving? Results from Constructed-Response Questions in NAEP's 1992 Mathematics Assessment*, 1993.

Dossey, Mullis, and Jones as providing evidence that "students understood the mathematics in the question, but failed to heed the instructions asking that work be shown, reasoning explained, and multiple budget options explored" (1993, p. 122). Thirty-seven percent provided minimal responses characterized by the presence of one or more calculations based on information given in the problem but that had no direct relationship to the question itself (identifying Treena's budget options) or that involved unrelated information. Finally, more than one-fifth of the students left their papers blank.

To extend what is known about student performance on the Treena's Budget question, especially the information reported by Dossey, Mullis, and Jones (1993, pp. 116–23), a sample of responses, representing about 30 percent of the total number of nonblank responses, was obtained and analyzed for this chapter. The intent of this analysis was to provide additional examples of the budget alternatives identified (or not identified) by students and how they communicated their solutions to the reader of their work, as specified in the directions to the problem ("Your answer should be clear enough so that another person could read it and understand your thinking"). Figure 5.2 contains these additional examples.

Response 1

train $125 ⎫
group $280 ⎬ could afford
food $315 ⎭   Total $720

Air $335 ⎫
group $280 ⎬ could afford
food $315 ⎭   Total $930

train $125 ⎫
individual $420 ⎬ Total $860 could afford
food $315 ⎭

Response 2

1000

335   125
420 280   420 280
315   315   315   315
(1070)   (930)   (860)   (720)

She could go by air & take group instruction, go by train & take individual instruction, or go by train & take group instruction.

Response 3

```
      $ 60
      ×  ? 1 week
$ 40   420
× ?
280    food
       $45.00
          7
      315.00
```

she could go by air $3350⁰
group instruction $ 280 ⁰⁰
$40 per day /week $ 315 00
food $45 per day $930.00
  / week is

Response 4

Have individual instruction
and take the train

Have group instruction take the train

Have groue instruction take the
plane

Response 5

She would go by train. Take the group
lessons and $45 a day for food. If she
doesn't want to spend a lot of money,
she should take the cheaper choices.

Fig. 5.2 Treena's Budget: Additional student responses

Responses 1 and 2 in figure 5.2 illustrate complete answers presented clearly; response 2 is particularly interesting because of its identification of all four alternatives through the use of a tree diagram as an organizational tool. In response 3 only one option is identified with computation provided as supporting work. Responses 4 and 5 are typical of those that listed correct options without any supporting work. It may be that the presence of the calculator influenced students not to record their work, despite the explicit directions to give a clear explanation and show all work.

From results reported by NAEP for Treena's Budget and from the analysis of a subset of student responses to that question, it seems clear that, at least in 1992, students had not had much experience with extended-response questions or with mathematical tasks that require explanation of concepts, reasons for computational steps, or multiple solutions. Helping students achieve a high level of mathematical communication may require changing students', teachers', and other educators' beliefs about the nature of mathematical evidence and the form of discourse in the mathematics classroom. Fitzgerald and Bouck (1993) offer rather broad suggestions to begin to change the mode of instruction in mathematics away from direct instruction to the use of mathematics activities and investigations, cooperative learning, and mathematics as inquiry. Koehler and Prior (1993) offer detailed suggestions for improving the nature of classroom interactions in mathematics, away from low-level teacher-initiated interactions to higher-level student-initiated and student-to-student interactions. However, the area of students' mathematical explanations remains one in need of further research and well-grounded practical approaches.

RATIONAL NUMBERS

In the content area of numbers and operations, about the same number of NAEP items involved rational numbers as involved whole numbers. However, more of the rational-numbers items were administered to students in grades 8 and 12 than to students in grade 4, a pattern that is reasonable given the school mathematics curriculum in this country. The next sections of this chapter present results on the NAEP rational-number concepts and properties, including representing, comparing, and rounding rational numbers, and rational-number operations and applications.

Rational-Number Concepts and Properties

The 1992 NAEP assessment included eighteen items that addressed the representation and comparison of rational numbers. Half the items involved fractions only, five items were exclusively decimal items, three items addressed fraction-decimal relationships, and one involved percent. Areas assessed in the items included equivalent representations of rational numbers (e.g., pictures of regions, decimal notation, word names), comparing rational numbers, and rounding.

HIGHLIGHTS—RATIONAL NUMBERS

- Students were generally successful in choosing pictorial representations for simple fractions, but they had more difficulty with more-complex representations involving equivalent fractions.
- Interpreting fractions and decimals as locations on a number line was more difficult for students than interpreting fractions as part of geometric regions.
- Both eighth- and twelfth-grade students were more successful on word problems involving multiplication of rational numbers than on division of rational numbers, whereas fourth-grade students had low levels of performance on both.
- Performance was better on multiplication involving at least one whole-number factor than on multiplication involving only rational numbers as factors.
- Many students, particularly those in grade 4, incorrectly interpreted multistep rational-number word problems as one-operation procedures.

Representing Rational Numbers

There were two types of items that involved representing rational numbers: those involving pictures and those involving symbols. A subset of seven items that appeared in NAEP used pictures of geometric regions and number lines; two used only fraction notation, decimal notation, or simply the word name for the rational number. For the pictorial representations, each item required a part-whole interpretation of rational numbers that, as described by Behr and Post (1992), "depends directly on the ability to partition either a continuous quantity or a set of discrete objects into equal-sized subparts or sets" (p. 203). No items required interpreting rational numbers as part of a set of discrete objects, but the geometric-region and number-line items required interpreting rational numbers as partitioning of a continuous quantity.

The set of items described in table 5.13 shows patterns of performance on items in which rational numbers were represented as parts of geometric regions or on number lines. Of special interest were those items that were administered at more than one grade, which afforded the opportunity to examine student performance across grade levels. For item 1 in the table, the majority of students at all three grade levels chose the correct picture of a region for a simple fraction. Performance increased sharply from grade 4 to grade 8—69 percent to 90 percent, respectively—and then remained

Table 5.13 Performance on Items Involving Representations of Rational Numbers

Item Description	Percent Correct		
	Grade 4	Grade 8	Grade 12
Geometric Region			
1. Choose a picture that represents a given fraction.	69	90	91
2. Choose a picture that represents an equivalent fraction.	42	70	—
3. Shade a fraction of a given rectangular region using an equivalent fraction.	18	63	—
4. Choose an equivalent fraction for a given picture.	—	67	—
Number Line			
5. Identify a number associated with a given point. (scale indicated in decimals)	39	84	—
6. Locate a point on a number line. (scale indicated in fractions)	27	58	77
7. Locate a point that represents 1.75. (scale indicated in fractions)	—	—	50

Note: Items 1, 2, and 4 were multiple-choice items, and the rest were regular constructed-response items.

essentially the same from grade 8 to grade 12. The growth between grades 4 and 8 most likely reflects the curricular emphasis on fractions in the middle grades. Notable was the less successful performance on the examples that dealt with equivalent fractions, as illustrated by items 2, 3, and 4. This is consistent with Behr et al.'s (1983) reported difficulty that students have when a geometric region needs to be interpreted visually in more than one way. For example, a region divided into eighths can be reinterpreted as fourths if certain partitions can be visually ignored. The very poor performance on item 3 by fourth-grade students is very likely due to lack of curricular experience with equivalent fractions, coupled with the fact that this item required a constructed response. While performance was also lower for eighth-grade students on items 3 and 4, the difference compared to item 2 was not as drastic as it was for the fourth grade, which perhaps reflects the middle grades' emphasis on rational numbers.

Items 5, 6, and 7 were each based on a picture of a number line. Performance on these items was lower than that of the first set, which involved geometric regions. Items involving number-line representations for rational numbers often have a number of features that may add to difficulty in interpretation. For example, Novillis-Larson (1980) found that locating fractions on number lines longer than one unit is a source of difficulty for students, and items 5 and 7 both have this feature. Additionally, she found that students are more successful locating fractions on number lines in which the number of segments is equal to the denominator of the fraction, compared to a situation in which the number of segments is double that of the denominator in the fraction, as in item 6. Another source of difficulty may have been the mixed use of decimal and fraction notation in this item. Markovits and Sowder (1991) (as reported by Sowder [1995]) found in a 1991 study that most middle-grades children were unsuccessful with a set of tasks that mixed fraction and decimal notation. Item 7, shown in detail in table 5.14, was answered correctly by only half of the twelfth-grade students. The item had at least three features that may have made it especially difficult: It used a mix of decimal and fraction notation in the prompt, it involved a number line that was two units long, and it also had three segments per unit, while asking the student to locate a decimal with an understood denominator of 100.

Of the two items that used only symbolic representation (i.e., fraction notation, decimal notation, or word names for the rational number), one

Table 5.14   Locating a Point on a Number Line: Sample Item

Item	Percent Responding Grade 12
On the number line below, place a dot at the point that could represent 1.75.	

$$0 \quad \frac{1}{3} \quad \frac{2}{3} \quad 1 \quad \frac{4}{3} \quad \frac{5}{3} \quad 2$$

Item	Percent Responding Grade 12
Correct response: Dot placed anywhere in the interval from $\frac{5}{3}$ to 2, but not on $\frac{5}{3}$ or 2.	50
Incorrect response: Dot placed at either $\frac{4}{3}$ or $\frac{5}{3}$.	33
Incorrect response: Dot placed between 1 and $\frac{4}{3}$.	9
Any other incorrect response other than the two types just described.	7
Omitted	1

released item that linked decimal notation to the word name for the rational number was revealing. The item, which was administered to both fourth- and eighth-grade students, assessed their ability to identify the decimal notation for a numeral given in words: "What number is four hundred five and three-tenths?" Students in those grades had little difficulty choosing the correct answer from a list of choices: Almost 70 percent of fourth-grade students and 93 percent of eighth-grade students selected 405.3 as the correct decimal notation. In this item zero as a place holder is located in the whole-number portion of the decimal numeral. Performance might not have been as high if the item instead had asked students to select the decimal representation in an example where the place holder appeared in the decimal portion of the notation, for example, "forty-seven and five hundredths."

Comparing Rational Numbers

When rational numbers are compared in a context, one considers whether the size of the whole unit is the same for the numbers being compared. There is evidence from the NAEP examples that some students may have difficulty understanding how to interpret the whole unit when comparing rational numbers. For example, an extended constructed-response item, commonly referred to as Pizza Comparison and shown in table 5.15, asked students in grade 4 to decide whether half of one pizza would be the same size as half of another pizza and then to explain their reasoning using words or pictures. The intent of the item is to measure students' ability to communicate mathematically that a fraction must be interpreted relative to the size of the object, which, in this case, was the size of the "whole" pizza.

Results for Pizza Comparison based on the five performance levels for extended constructed-response questions appear in table 5.15; descriptions of the performance levels and sample responses for each level are shown in figure 5.3. The descriptions and sample responses are as they appeared in an NAEP publication (Dossey, Mullis, and Jones 1993, pp. 93–95). About one-fourth of the fourth-grade students produced either extended or satisfactory responses that communicated the relationship between fractional part and relative size. Another 18 percent of the students declared that Ella was right, and their reasoning involved the misconception that "1/2 always equals 1/2." There is some indication that this question was difficult for fourth-grade students: 7 percent did not attempt to answer it and another 49 percent did not use fraction concepts in communicating their responses.

As was the case for the extended constructed-response question Treena's Budget described previously in this chapter, the Pizza Comparison problem was described in Dossey, Mullis, and Jones (1993), but the discussion was almost exclusively limited to explanation of the score levels. To gain a better understanding of the kinds of responses students gave to this extended problem, a sample representing about 35 percent of the total num-

Table 5.15 Pizza Comparison

Question	Percent Responding Grade 4
[General directions]	
Think carefully about the following question. Write a complete answer. You may use drawings, words, and numbers to explain your answer. Be sure to show all of your work.	
José ate $\frac{1}{2}$ of a pizza.	
Ella ate $\frac{1}{2}$ of another pizza.	
José said that he ate more pizza than Ella, but Ella said they both ate the same amount. Use words and pictures to show that José could be right.	
Extended response	16
Satisfactory response	8
Partial response	2
Minimal response	18
Incorrect	49
Omitted	7

ber of nonblank responses was obtained and analyzed for this chapter. The intent of the analysis of the sample responses was to gain additional information on how students justified the premise that José could be right or how they supported the notion that Ella was right. Figure 5.4 contains examples of responses based on José's being right or Ella's being right.

For responses in which José was selected as being right, the explanations were most often based on the notion that José's pizza was bigger than Ella's pizza. Most of those responses were expressed in both words and pictures, and the example of an "extended" response in figure 5.3 is a typical example of that kind of response. Responses 1 and 2 in figure 5.4 show other ways in which students justified José's being right, with response 1 using labeled pictures and response 2 using only words. However, some responses had explanations about why José could be right that were based on introducing new constraints into the problem or changed the existing

Incorrect—The work is completely incorrect or irrelevant, or the response states, "I don't know."

José ate ½

Minimal—Student responds that "1/2 is always 1/2" indicating an awareness of fractional parts. Other responses may include only references to number of pizzas or to toppings.

Jose's half   Ella's half

Jose ate his ½ and Ella ate her ½ they both had ½ and they both ate the same amount.

Partial—Student makes statements such as "José pizza has bigger pieces" that begin to demonstrate an awareness of the idea of relative size.

$\frac{1}{2}$ ⊗ Jose could be right because it is a maybe bigger pise of pizza.

Satisfactory—Student displays responses that connect figurally the relationship between the difference in the relative size of José's and Ella's pizzas but are not clear in explaining that relationship.

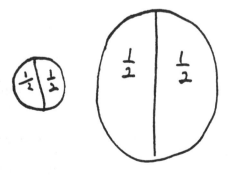

Continued

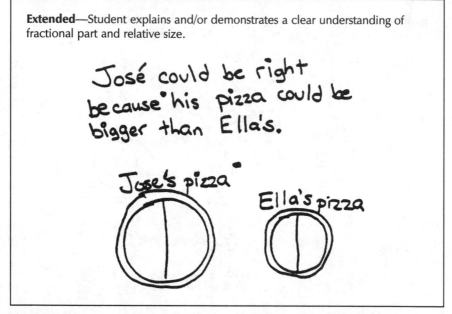

Fig. 5.3 Pizza Comparison: Performance categories and sample responses
Source: Dossey, Mullis, and Jones. *Can Students Do Mathematical Problem Solving? Results from Constructed-Response Questions in NAEP's 1992 Mathematics Assessment,* 1993.

constraints. For example, some students wrote that José could have eaten *more* than half of his pizza, which thus changed the constraint that they both ate half of their respective pizzas. Response 3 in the figure is representative of a situation that occurred in a number of responses in the sample, showing that students used their real-world knowledge of pizza. Rather than thinking mathematically about fractions, those students relied on ideas like José's pizza had more toppings than Ella's, the crust on his pizza was thicker than the crust on her pizza, or that he ate the crust and Ella did not eat this part of her pizza.

Of the sample responses that disagreed with the premise that José could be right, many responses directly stated or indirectly implied the expected misconception: "José cannot be right because 1/2 always equals 1/2." Response 4 in figure 5.4 is an example of an implied use of that misconception. Here, the student drew two equal-sized pizzas, cut each in half, and concluded that "they ate the same amount," which thus indirectly implied the misconception just stated. Interestingly, though, there was evidence that students assumed that the two children were sharing one pizza. Responses 5 and 6 in figure 5.4 are examples of this situation that suggest that

Response 3

José may be rite
because he might
have ate more peperonie
on his pizza

Response 2

maybe the
other pizza was
smaller than
José,

Response 1

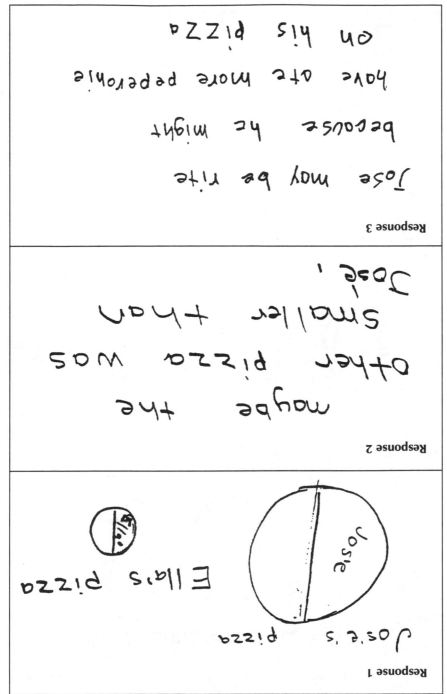

José's pizza   Ella's pizza

Josie

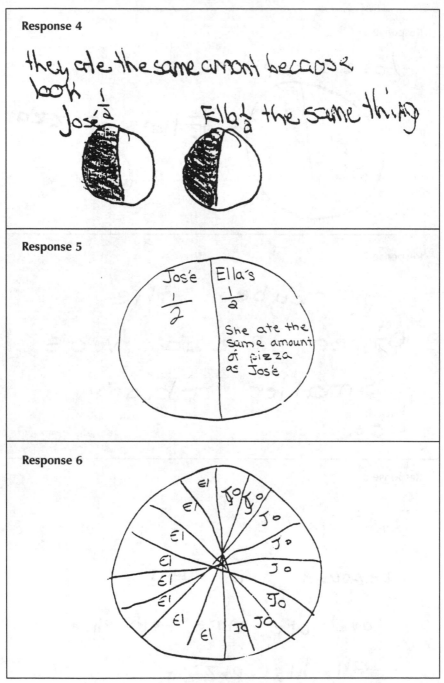

Response 4

they ate the same amont becaose
book
Jose $\frac{1}{2}$ Ella $\frac{1}{2}$ the same thing

Response 5

José   Ella's
$\frac{1}{2}$ $\frac{1}{2}$

She ate the
same amount
of pizza
as José

Response 6

El El Jo Jo
El Jo
El Jo
El Jo
El Jo
El Jo
El Jo Jo

Fig. 5.4 Pizza Comparison: Additional student responses

students may have misread the problem. In their analysis of the items on the 1992 fourth-grade NAEP test (Silver and Kenney 1994), a panel of experts commented that the wording used in the Pizza Comparison problem involves a rather subtle use of the word *another* to imply that each child had his or her own pizza. If a student did not read carefully, then the existence of one pizza for each child might be missed, which would increase the chance that a choice of an incorrect justification would be due to reading comprehension rather than to a mathematical misconception.

The patterns identified in the sample of responses obtained and analyzed for this chapter are similar to those reported by NAEP. Many students disagreed with the premise in the Pizza Comparison problem that José could be right, believing instead that Ella is right and not realizing the potential for the two quantities being compared to be different sizes. Thus, some fourth-grade students may be confused about the role of the unit whole for this situation. However, it is encouraging that there is evidence both from NAEP results from the focused holistic scoring scheme and from the qualitative analysis of responses done for this chapter that some fourth-grade students acknowledged the relevance of size of the whole unit in comparing fractions. This encouraging level of performance by students prior to a heavy curricular emphasis on rational numbers may be explained by Mack's (1990) research that indicates that many students possess knowledge related to the mathematical idea of the relationship between the number of parts in a whole and the size of the parts, especially for real-world problems (such as pizza) and for the fraction 1/2. Also, it was evident from the responses in the sample that some students made their own personal sense of this problem not through the use of mathematics but rather through the use of their real-world experiences with eating pizza and that other students may have misread the problem and missed the fact that each child had a pizza.

A regular constructed-response item, given to twelfth-grade students and shown in table 5.16, illustrates that some older students also have trouble identifying and interpreting the "whole," in this case when comparing percents. For this item, the rational numbers require part-whole interpretations for two different sets of voters. A correct response to the item involves identifying that the total number of votes for females and males is missing, which makes it impossible to determine which candidate got more votes. Over half the twelfth-grade students responded incorrectly to this problem, which targeted a problem in understanding the role of the "whole." Another 17 percent chose the option Cannot Tell but either gave an incorrect explanation or no explanation. In addition to the general difficulty students had in explaining their answers across all NAEP constructed-response items, it may be that requiring students to interpret two different and unknown "wholes" in a problem further increased the difficulty of the item.

Table 5.16 Comparing Percents: Sample Item

Item	Percent Responding Grade 12
In a recent election between Candidates A and B, 48 percent of the male voters and 55 percent of the female voters voted for Candidate A. Did Candidate A receive more votes than Candidate B?	
○ YES	
○ NO	
○ CANNOT TELL	
Explain your answer.	
Cannot tell: need to know the number of males and the number of females	26
Cannot tell with an incorrect explanation or no explanation	17
Incorrect answer (Yes or No) with or without an explanation.	52
Omitted	5

The ability to compare rational numbers goes beyond interpreting the whole to also understanding the relationship between the denominator (i.e., the size of the different individual parts) and the numerator (i.e., the number of those individual parts). For example, one such NAEP item required a regular constructed response in a real-world context. Fourth-grade students were to compare two simple fractions, each with the numeral 1 in the numerator, and to draw a picture or use words to explain their answer. Only 12 percent of the fourth-grade students answered correctly. The unsuccessful performance may have resulted from difficulty with interpreting the denominators or because the item required a constructed response. In another comparison item, eighth-grade students were to select a set of three fractions arranged from least to greatest; only one fraction had the numeral 1 in the numerator. Only 30 percent of the eighth-grade students selected the correct choice. Even more students, 33 percent, selected a distracter where the denominators were arranged from greatest to least. Considering that some of the students were likely to be guessing randomly, at least some of the 33 percent may not be considering the size of the fractional pieces as indicated by the denominator.

Rounding Rational Numbers

Rounding numbers is a traditional topic in the elementary and middle school curriculum and is important for estimating the results of calculations. Further, students need to deal with rounding in situations where calculators display results that go beyond the size of the display panel, and when calculating with measurement in order to have the results be meaningful in the context of the real world. The items shown in table 5.17 illustrate a variety of tasks that address rounding rational numbers. Interestingly, only one of the three items uses the word *round* in the prompt. Instead, items 1 and 3 give the value to which the number is (implicitly) rounded and asks students to select which of the choices in the set is "closest." Item 2 embeds the rounding in a computational estimation situation. A fourth item that was not released to the public directly asks students to round an amount of money, which resulted in a high level of performance by fourth-grade students, with 87 percent selecting the correct answer. Performance on item 1 was not as successful for fourth-grade students, which may be due to the different format (as discussed above) but also the fact that students are less familiar with "seconds" than they are with money. However, eighth-grade students were very successful, which perhaps reflects the curricular emphasis on rational numbers during the middle grades. Fourth-grade performance on item 2 was very low, perhaps because there were four rounding tasks embedded in the prompt. Eighth-grade performance was also lower on this item than on item 1, but the difference was not as drastic as that of the fourth grade. Coupled with the very successful twelfth-grade performance, a reasonable conclusion may be that some students are learning to round successfully over the grades. Item 3 was answered correctly by only half of the eighth-grade students. Because this item included a mix of decimal and fraction representations, it may be that the low level of performance was due to difficulties students have in tasks that have mixed fraction and decimal notation, as was noted in the discussion on number-line items above.

Rational-Number Operations and Applications

The NAEP assessment included thirteen items dealing with multiplication or division of rational numbers, nine in the context of word problems and four in abstract form. No items were included that focused exclusively on addition or subtraction; however, four word problems involved a combination of all four operations.

Table 5.17 Rounding Rational Numbers: Sample Items

Item	Percent Responding		
	Grade 4	Grade 8	Grade 12
1. Which of the following is closest to 15 seconds?			
A. 14.1 seconds	16	2	—
B. 14.7 seconds	2	<1	—
C. 14.9 seconds*	63	92	—
D. 15.2 seconds	18	5	—

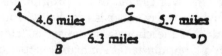

Item	Grade 4	Grade 8	Grade 12
2. Carol wanted to estimate the distance from A to D along the path shown on the map above. She correctly rounded each of the given distances to the nearest mile and then added them. Which of the following sums could be hers?			
A. $4 + 6 + 5 = 15$	36	10	3
B. $5 + 6 + 5 = 16$	13	8	3
C. $5 + 6 + 6 = 17$*	25	75	89
D. $5 + 7 + 6 = 18$	17	6	4
3. Of the following, which is closest in value to 0.52?			
A. $\frac{1}{50}$	—	29	—
B. $\frac{1}{5}$	—	11	—
C. $\frac{1}{4}$	—	6	—
D. $\frac{1}{3}$	—	4	—
E. $\frac{1}{2}$*	—	51	—

*Indicates correct response.

Note: Percents may not add to 100 because of rounding or omissions.

Multiplication and Division

The multiplication and division items were of two types. One type involved abstract multiplication and division, and the other involved word problems placed in real-world contexts. To NAEP's credit, among the four abstract multiplication and division items were some that involved straightforward computation and others that probed concepts underlying the meaning and use of the operations. On a multiplication item involving decimal factors, almost two-thirds of fourth-grade students found the correct product. This was a calculator item, and as might be expected, those reporting calculator use were more successful than those who reported not using the tool: 80 percent compared to 40 percent. On a division item where the quotient to a whole-number division problem needed to be expressed as a decimal, fourth-grade students were fairly successful, with 66 percent selecting the correct decimal quotient. Sixty-eight percent of those reporting the use of a calculator responded correctly, and a surprisingly high 58 percent of non–calculator users were successful. It may be that students could readily rule out two of the distracters using estimation of magnitude, inflating the level of successful responses based on guessing. Less successful were twelfth-grade students who divided 30.45 by 15. Only 68 percent produced the quotient, 2.03, but another 20 percent of the students calculated an answer of 2.3. Much as in the discussion on the link between decimal notation and word names for rational numbers, this result may indicate difficulty students have with 0 as a place holder in the decimal portion of the quotient.

A constructed-response item displayed in table 5.18 is especially revealing of students' knowledge of multiplication with rational numbers because it probes an assumption that changes as one moves from a primarily counting-number interpretation to rational-number interpretation of multiplication. Of the eighth-grade students, 58 percent reported that Tracy was correct, although only 48 percent also provided a correct explanation. Twelfth-grade performance was more successful, with 63 percent indicating that Tracy was correct along with a sound explanation and an additional 13 percent also selecting Tracy but without a supporting explanation.

NAEP's scoring and reporting methods did not provide any information on the nature of the reasons students gave for Tracy's being correct. Of particular interest would be the kinds of numbers that students gave as examples for factors that, when coupled with 6, would produce a product less than 1 (for example, 0, numbers between 0 and 1, negative numbers). To gain additional information about the kinds of numbers given as examples and whether these examples changed from grade 8 to grade 12, samples of responses were obtained for each grade level (849 responses from eighth-grade students and 576 for twelfth-grade students) and then analyzed according to the examples given to show that Tracy was right. In

Table 5.18 Multiplying Rational Numbers: Sample Item

	Percent Responding	
Item	Grade 8	Grade 12
1. Tracy said, "I can multiply 6 by another number and get an answer that is smaller than 6." Pat said, "No, you can't. Multiplying 6 by another number always makes the answer 6 or larger." Who is correct? Give a reason for your answer.		
Tracy, with correct reason given	48	63
Tracy, with no reason or incorrect reason	10	13
Any response that states or suggests that Pat is correct.	38	23
Omitted	3	1

Note: Percents may not add to 100 because of rounding.

both grades, the most frequent example found in the set of sample responses was 0, given by 45 percent and 48 percent of the eighth- and twelfth-grade students, respectively. A typical response was "Tracy is right because $6 \times 0 = 0$ and zero is a smaller number than 6." Seventeen percent of the eighth-grade students in the sample gave examples based on negative numbers, and this percent increased to 34 percent for students in grade 12. Similar patterns of increased percents from grade 8 to grade 12 were found in the sample for examples based on numbers between 0 and 1 (13 percent in grade 8 and 28 percent in grade 12). Additionally, more twelfth-grade students than eighth-grade students (13 percent and 7 percent, respectively) gave more than one numerical example to show that Tracy was right (for example, "Tracy, because she could use anything below 1—like 0.5 or –6"). This pattern of responses suggests that high school students may have a more developed sense of rational numbers, especially decimals and fractions, than eighth-grade students.

The second set of multiplication and division items consisted of word problems set in a context. Of the nine items, five involved multiplication and four involved division. Table 5.19 contains a summary of these items. With the exception of item 1, an item for which students were permitted to use a calculator, the performance of fourth-grade students was quite low, which is not unexpected, considering that more curricular emphasis is given to operating with rational numbers in the middle school years than in the elementary school grades. Twelfth-grade students' performance was relatively high for three of the four items administered at that grade level.

Eighth-grade students were administered the bulk of the multiplication and division word problems. Those students were most successful on rational-number multiplication items that involved one whole-number factor, with 82 percent, 64 percent, and 73 percent answering correctly on items

Table 5.19   Performance on Rational-Number Multiplication and Division Word Problems

Item Description	Percent Correct		
	Grade 4	Grade 8	Grade 12
Multiplication			
1. Select the cost for a large number of eggs.	60	82	—
2. Select number of cups of flour needed for multiple batches of cookies based on the amount of flour needed for each batch.	21	64	—
3. Calculate the amount of money collected from the sale of movie tickets.	—	73	—
4. Given the rate of 0.6 miles per day, calculate the number of miles run in 45 days.	—	—	89
5. Find a fraction of a fraction.	—	—	20
Division			
6. Choose correct answer to a word problem involving division of fractions.	24	52	69
7. Find the quotient for a word problem involving a whole number and a fraction, and choose the correct answer.	—	53	—
8. Write a word problem involving division of a whole number and a fraction.	—	25	—
9. Calculate the number of 85-cent rulers that can be purchased with $7.00.	—	—	75

Note: Items 3, 4, 5, 8, and 9 were regular constructed-response items, and the rest were multiple-choice items.

1, 2, and 3, respectively. This result is consistent with Harel and Confrey's (1994) finding (as reported in Harel [1995]) that student performance is higher when the multiplier is a whole number than when the multiplier is a decimal. In such items, multiplication can be conceptualized as repeated addition, just as it is with whole-number multiplication, whereas finding a fraction of a fraction is a different conceptualization of multiplication. Twelfth-grade students' poor performance on item 5 may indicate that this shift in conceptualizing multiplication is very difficult even for high school students.

The three division items administered to the eighth-grade students present a somewhat troubling picture. Only about half of the students responded correctly to multiple-choice items 6 and 7, and only 25 percent responded correctly to item 8, a regular constructed-response item. Given that these division items could be solved using a variety of methods, such as drawing pictures, reasoning about the situation, or using the standard division algorithm, it is disturbing that eighth-grade students apparently do not have enough conceptual grounding in rational numbers and the operation of division for a higher level of success on these items.

Rational-Number Word Problems Using Combinations of Operations

Five items from the NAEP assessment required students to use different combinations of operations for problems involving money. Typically, items were similar to the one in table 5.20. There was evidence that some students apparently incorrectly reduced the item to a one-operation procedure. For example, while 21 percent of fourth-grade students responded

Table 5.20   Rational-Number Word Problem

Item	Percent Responding Grade 4
George buys two calculators that cost $3.29 each. If there is no tax, how much change will he receive from a $10 bill? Answer: _____	
Correct response of $3.42	21
Incorrect response of $6.71	28
Any other incorrect response	48
Omitted	3

correctly to the item in table 5.20, another 28 percent responded with $6.71, which indicates that students subtracted the cost of one calculator from $10. A similar phenomenon was detected in two other secure items administered to fourth-grade students. On these two items, 23 percent and 30 percent of fourth-grade students selected the correct answer, while 49 percent and 26 percent chose an answer involving incorrectly reducing the problem to a one-operation procedure. The second of the two items was also administered to eighth- and twelfth-grade students, which provides some evidence that this phenomenon may decrease drastically over the grades. Of the eighth-grade students, 74 percent answered correctly and 7 percent selected the distracter that was consistent with incorrectly reducing the problem to a one-operation procedure. Lastly, for twelfth-grade students the percents associated with the correct answer and the incorrect distracter were 86 percent and 3 percent, respectively.

Percent, Ratio, and Rate Word Problems

The NAEP assessment included items on percent, ratio, and rate. These topics typically are formally introduced in the middle grades and revisited throughout the secondary curriculum as students encounter topics such as similarity and direct variation. Additionally, students in elementary school often encounter ratio and rate in the context of multioperation, whole-number word problems.

Three percent items involving exact or estimated calculations with percents were included in the NAEP assessment. Two of the items were administered to students in grade 8, and one overlap item, which is shown in table 5.21, was administered to students in grades 8 and 12. Performance by eighth-grade students was low on these three items, ranging from 33 to 40 percent correct. The item in the table indicates better performance by twelfth-grade students. Of particular interest is that 25 percent of eighth- and 20 percent of twelfth-grade students selected an answer of $806, which is an incomplete response reflecting the tax and extra fees, not the final amount paid. Overall, the performance on items involving calculations with percent is disappointing.

Seven items in the NAEP assessment set involved rates, ratio, or both. Of the seven items, three were administered only to fourth-grade students and three only to twelfth-grade students. One item was administered across all three grade levels, which provided some information about growth over the grades. Since ratios and rates involve composite units, students often approach these types of problems as multistep, whole-number operation items. Notable in the fourth-grade performance was the predominance of incorrect choices that were consistent with reducing the problem to a one-operation procedure. For example, in table 5.22, item 1 displays the most difficult of the fourth-grade items. Of the fourth-grade students, 59 percent

Table 5.21   Word Problem for Estimating with Percent

Item	Percent Responding	
	Grade 8	Grade 12
Ken bought a used car for $5,375. He had to pay an additional 15 percent of the purchase price to cover both the sales tax and extra fees. Of the following, which is closest to the <u>total</u> amount Ken paid?		
A. $806	25	20
B. $5,510	14	4
C. $5,760	12	3
D. $5,940	7	3
E. $6,180*	40	69

*Indicates correct response.

Note: Percents may not add to 100 because of rounding or omissions.

selected a response of $1.14, which is the difference of the two packages of birdseed. This answer disregards the relationship between the price and the size of the packages and simply finds the difference of the listed prices.

While incorrectly reducing multiple-step problems to one-operation procedures for these problems was similar to the phenomenon noted above in the mixed-operations word problems, an examination of responses by the twelfth-grade students did not indicate that older students were also responding this way. However, studying popular distracters provided some indication that a number of students may have been correctly implementing multiple operations but were inclined to stop before completing all the required operations. For example, in item 2 in table 5.22, while only 5 percent of the students responded correctly to the item, 11 and 10 percent gave correct answers to parts A and B, respectively, although not to both.

Growth over the grades is difficult to ascertain because the range in percent-correct values for the three items administered only to fourth-grade students was 8 percent to 58 percent, and for the three other items administered only to twelfth-grade students, it was 5 percent to 50 percent. Interpretation is difficult partly because of the greater difficulty of the problems given to twelfth-grade students. However, one item was administered across grades and required students to interpret and use a scale. Successful performance on the multiple-choice item was achieved by 20 percent of fourth-grade students, 36 percent of eighth-grade students, and 50 percent of twelfth-grade students. Whereas growth is evident over the grades, the overall low level of performance on percent, ratio, and rate word problems is troubling.

Table 5.22 Word Problems Using Rates, Ratios, and Proportions: Grade 4 and Grade 12

	Percent Responding	
Item	Grade 4	Grade 12

1. A package of birdseed costs $2.58 for 2 pounds. A package of sunflower seeds costs $3.72 for 3 pounds. What is the difference in the cost <u>per pound</u>?

	Grade 4	Grade 12
A. $0.05*	8	—
B. $1.14	59	—
C. $1.24	14	—
D. $1.29	17	—

Video Store A	Video Store B
$2.65 per tape for one night	$3.00 per tape for 2 nights
$1.50 charge for each additional night	1 credit if tape returned after one night
Every 10th tape **free** for one night	Every 10 credits = one **free** rental

2. The Peterson family rents 30 videotapes yearly, of which 23 are rented for one night only and 7 are rented over a period of two nights. Given the rental fee structures shown above, fill in the chart below with the total yearly costs for the Petersons at each store. (Note: The 30 tapes include the free tapes earned.)

Store	Total Cost
A	
B	

	Grade 4	Grade 12
Correct answer: $82.05 for Video Store A and $84.00 for Video Store B	—	5
Correct answer for Video Store A; incorrect answer for Video Store B	—	11
Correct answer for Video Store B; incorrect answer for Video Store A	—	10
Any other incorrect response	—	71
Omitted	—	3

*Indicates correct response.

Note: Percents may not add to 100 because of rounding or omissions.

OVERALL PERFORMANCE IN NUMBERS AND OPERATIONS

Table 5.23 shows average performance across the entire set of numbers and operations items in 1990 and 1992 in terms of the NAEP scale and according to overall performance and performance by gender and race/ethnicity. At grades 4, 8, and 12, overall performance in numbers and operations increased significantly from 1990 to 1992. In both years there was a greater increase in performance in numbers and operations between grade 4 and grade 8 than between grade 8 and grade 12. This steeper increase between the elementary grades and the middle grades is not surprising, given the focus on numbers and operations that is typical of the elementary mathematics curriculum. The emphasis placed on number concepts is reflected in results from the questionnaires completed by teachers of the fourth- and eighth-grade students who took NAEP in 1992. Teachers' responses to a question about how much emphasis they gave (or will give) to the topic of numbers and operations showed that 92 percent of students in grade 4 had teachers who reported giving heavy emphasis to that topic, but only 76 percent of students in grade 8 had teachers who reported doing so.

Both males and females significantly increased their performance levels in numbers and operations from 1990 to 1992. There appeared to be no gender differences either in 1990 or 1992 at any grade level. For the race/ethnicity subgroups, White students in all three grades and Black and Hispanic students at grade 12 had significant performance gains from 1990 to 1992. In 1992, there were disparities in performance in numbers and operations between race/ethnicity subgroups. White students scored at significantly higher levels than both Black and Hispanic students at all three grade levels, and Hispanic students scored at significantly higher levels than Black students in grades 4 and 8.

CONCLUSION

The NCTM *Curriculum and Evaluation Standards for School Mathematics* (1989) includes standards related to numbers and operations as part of the Communication and Problem Solving standards at all three grade-level ranges (K–4, 5–8, and 9–12) and in the other standards at the two lower levels: the Number Sense and Numeration, Concepts of Whole Numbers, and Whole Number Computation standards at grades K–4 (pp. 38–47) and the Number and Number Relationships, Number Systems and Number Theory, and Computation and Estimation standards at grades 5–8 (pp. 87–93). Although not as extensive as the NCTM *Standards,* the 1992 NAEP objectives emphasize many of the same concepts and skills as the *Standards.*

Table 5.23 Average Performance in Numbers and Operations: Overall Performance and Performance by Gender and Race/Ethnicity, Grades 4, 8, and 12 for 1990 and 1992

	Grade 4	Grade 8	Grade 12
Overall			
1992	216 (0.8)>	272 (0.8)>	298 (0.9)>
1990	210 (1.1)	267 (1.3)	293 (1.1)
Gender			
Male			
1992	217 (0.9)>	271 (1.0)>	299 (1.0)>
1990	210 (1.4)	266 (1.6)	296 (1.3)
Female			
1992	214 (1.2)>	272 (1.0)>	297 (1.0)>
1990	210 (1.3)	267 (1.3)	290 (1.2)
Race/Ethnicity			
White			
1992	224 (1.0)>	280 (0.9)>	304 (0.9)>
1990	218 (1.3)	274 (1.3)	299 (1.2)
Black			
1992	189 (1.4)	244 (1.3)	276 (1.5)>
1990	187 (1.9)	246 (2.8)	271 (1.8)
Hispanic			
1992	198 (1.8)	251 (1.5)	282 (1.8)>
1990	195 (2.2)	249 (2.7)	275 (2.9)

> The value for 1992 was significantly higher than the value for 1990 at about the 95 percent confidence level. The standard errors of the estimated proficiencies appear in parentheses.

Results from the 1992 NAEP mathematics assessment reveal that students at all three grade levels appear to have an understanding of place value, rounding, and number theory concepts for whole and rational numbers in familiar, straightforward contexts. Students' understanding improves across grade levels but falls when the contexts are unfamiliar or complex.

Students at all three grade levels perform well on addition and subtraction word problems with whole and rational numbers that are set in familiar contexts and involve only one step or calculation. Eighth- and twelfth-grade students do well on multiplication and division problems involving one step, as long as one of the factors involved is a whole number. Difficulties arise when fourth-grade students apply an incorrect fall-back strategy of "when in doubt, add" and when some students at all three grade levels attempt to solve multistep problems as though they involved single-step procedures. It is unclear whether this incorrect strategy is due to a less-than-thoughtful approach to solving the problem, an inability to deal with more than one condition in a context, or poorly developed reading skills. More research is needed to clarify the source of difficulty and to recommend approaches for reducing the difficulties.

Additionally, with regard to rational numbers, students seem more familiar with and adept at using geometric area representations than number line representations, and some students have difficulty in applying the concept of a constant referent (whole unit) when comparing rational numbers. These difficulties may be linked to issues in instructional practice; however, continued in-depth research is needed.

The most troubling results were the low performance levels associated with students' ability to justify or explain their answers to regular and extended constructed-response items. Although in some cases performance did increase across grade levels, the percent of students responding correctly on the regular constructed-response items and at either the satisfactory or extended levels on the extended constructed-response questions remained low, often at less than 25 percent or even as low as less than 10 percent. The ability to communicate effectively about mathematics and in the language of mathematics is recognized in all the NCTM *Standards* documents—curriculum, teaching, and assessment—and in other education documents as increasingly important for all students now and in the future. The 1992 NAEP results suggest that in order to achieve the level of mathematical communication suggested as necessary for a mathematically literate citizen, students will need considerably more experience with justifying and explaining their mathematical work than they have had in the past. Concomitantly, research is needed in identifying both the nature of expectations regarding communication in mathematics and the means for achieving it in mathematics classrooms.

REFERENCES

Behr, Merlyn J., Richard A. Lesh, Thomas R. Post, and Edward A. Silver. "Rational Number Concepts." In *Acquisition of Mathematics Concepts and Processes*, edited by Richard Lesh and Marsha Landau, pp. 92–126. New York: Academic Press, 1983.

Behr, Merlyn J., and Thomas R. Post. "Teaching Rational Number and Decimal Concepts." In *Teaching Mathematics in Grades K–8: Research-Based Methods*, 2nd ed., edited by Thomas R. Post, pp. 201–48. Boston: Allyn & Bacon, 1992.

Dossey, John A., Ina V. S. Mullis, and Chancey O. Jones. *Can Students Do Mathematical Problem Solving? Results from Constructed-Response Questions in NAEP's 1992 Mathematics Assessment*. Washington, D.C.: National Center for Education Statistics, 1993.

Fitzgerald, William M., and Mary K. Bouck. "Models of Instruction." In *Research Ideas for the Classroom: Middle Grades Mathematics*, edited by Douglas T. Owens, pp. 244–58. New York: Macmillan, 1993.

Graeber, Anna O., and Elaine Tanenhaus. "Multiplication and Division: From Whole Numbers to Rational Numbers." In *Research Ideas for the Classroom: Middle Grades Mathematics*, edited by Douglas T. Owens, pp. 99–117. New York: Macmillan, 1993.

Harel, Guershon. "From Naive-Interpretist to Operation-Conserver." In *Providing a Foundation for Teaching Mathematics in the Middle Grades*, edited by Judith T. Sowder and Bonnie P. Schappelle, pp. 143–65. Albany, N.Y.: State University of New York (SUNY) Press, 1995.

Harel, Guershon, and Jere Confrey. *The Development of Multiplicative Reasoning in the Learning of Mathematics*. Albany, N.Y.: State University of New York (SUNY) Press, 1994.

Koehler, Mary Schatz, and Millie Prior. "Classroom Interactions: The Heartbeat of the Teaching/Learning Process." In *Research Ideas for the Classroom: Middle Grades Mathematics*, edited by Douglas T. Owens, pp. 280–98. New York: Macmillan, 1993.

Kroll, Diana Lambdin, and Tammy Miller. "Insights from Research on Mathematical Problem Solving in the Middle Grades." In *Research Ideas for the Classroom: Middle Grades Mathematics*, edited by Douglas T. Owens, pp. 58–77. New York: Macmillan, 1993.

Mack, Nancy K. "Learning Fractions with Understanding: Building on Informal Knowledge." *Journal for Research in Mathematics Education* 21 (January 1990): 16–32.

Markovits, Zvia, and Judith Sowder. "Students Understanding of the Relationship between Fractions and Decimals." *Focus on Learning Problems in Mathematics* 13 (January 1991): 3–11.

National Assessment of Educational Progress. *Mathematics Objectives: 1990 Assessment*. Princeton, N.J.: Educational Testing Service, National Assessment of Educational Progress, 1988.

———. *The Third National Mathematics Assessment: Results, Trends, and Issues*. Denver: Education Commission of the States, 1983.

National Council of Teachers of Mathematics. *Curriculum and Evaluation Standards for School Mathematics*. Reston, Va.: National Council of Teachers of Mathematics, 1989.

Novillis-Larson, Carol. "Locating Proper Fractions." *School Science and Mathematics* 53 (May 1980): 423–28.

Silver, Edward A., and Patricia Ann Kenney. "The Content and Curricular Validity of the 1992 NAEP TSA in Mathematics." In *The Trial State Assessment: Prospects and Realities: Background Studies*, pp. 231–84. Stanford, Calif.: National Academy of Education, 1994.

Silver, Edward A., Lora J. Shapiro, and Adam Deutsch. "Sense Making and the Solution of Division Problems Involving Remainders: An Examination of Middle School Students' Solution Processes and Their Interpretations of Solutions." *Journal for Research in Mathematics Education* 24 (March 1993): 117–35.

Sowder, Judith T. "Instructing for Rational Number Sense." In *Providing a Foundation for Teaching Mathematics in the Middle Grades*, pp. 15–29. Albany, N.Y.: State University of New York (SUNY) Press, 1995.

———. "Making Sense of Numbers in School Mathematics." In *Analysis of Arithmetic for Mathematics Teaching*, edited by Gaea Leinhardt, Ralph Putnam, and Rosemary A. Hattrup, pp. 1–51. Hillsdale, N.J.: Lawrence Erlbaum Associates, 1992.

Sowder, Judith T., and Judith Kelin. "Number Sense and Related Topics." In *Research Ideas for the Classroom: Middle Grades Mathematics*, edited by Douglas T. Owens, pp. 41–57. New York: Macmillan, 1993.

What Do Students Know
about Measurement?

Patricia Ann Kenney & Vicky L. Kouba

MEASUREMENT is one of the most practical areas of school mathematics because of its use in everyday life, in occupations, and in other areas of the school curriculum. This chapter provides a glimpse of what students know and can do in this content area assessed by NAEP. The first section describes measurement in NAEP and includes information about the set of measurement items on the 1992 assessment. Student performance data on individual measurement items and clusters of related items are reported in the next section. The final section presents information on overall performance on measurement in NAEP.

MEASUREMENT IN NAEP AND THE NAEP MEASUREMENT ITEMS

According to the NAEP framework, measurement "focuses on students' ability to describe real-world objects using numbers" (National Assessment of Educational Progress 1988, p. 21). The percent distribution of NAEP measurement questions recommended in the framework was 20 percent at grade 4 and 15 percent at grades 8 and 12. Topics to be assessed by NAEP included identifying attributes, selecting appropriate units of measure, reading and using measurement instruments, applying measurement concepts, and communicating measurement-related ideas. To assess students' understanding of perimeter, area, volume, and surface area, geometric figures were used, with appropriate figures selected for each grade level; for example, squares and rectangles at grade 4 and triangles and quadrilaterals at grades 8 and 12.

The 1992 NAEP mathematics assessment contained sixty-two measurement items, with a subset of those items administered at more than one grade level to facilitate comparisons of student performance among grade levels. More than 65 percent of the items were in multiple-choice format, and about 35 percent were constructed-response items that required students to produce their own answers or explanations. While working on certain measurement items, students were permitted to use calculators, paper rulers or combination ruler/protractors, or cardboard manipulatives in the form of geometric shapes. The sections that follow discuss student performance on individual NAEP measurement items and clusters of related items in these three categories: selecting, reading, and using measurement instruments; selecting and converting between units of measurement; and perimeter, area, volume, and surface area.

SELECTING, READING, AND USING MEASUREMENT INSTRUMENTS

Four secure multiple-choice items required students to select the appropriate measurement instrument for a particular situation. Each item was similar in structure to this example: "Which of the following would be used to find the length of your desk at school?" followed by a list of choices such

HIGHLIGHTS

- Students in grades 8 and 12 were able to read a variety of measurement instruments, but fourth-grade students have some difficulty reading measurement instruments when the object to be measured is not aligned with the beginning of the instrument and when the scale on the instrument involves an increment other than 1.
- Students at all three grade levels have a sense of appropriate units of measurement for particular situations and can make simple conversions from one measurement to another within the same system (customary, metric).
- Students at all three grade levels have some understanding of perimeter, and their understanding increases more sharply between grades 4 and 8 than between grades 8 and 12.
- Both fourth- and eighth-grade students show evidence of an incomplete conceptual understanding of area, sometimes confuse area and perimeter, and have difficulty applying area concepts to complex situations.
- Volume and surface-area concepts were difficult for twelfth-grade students.
- For students in all three grades, requiring written explanations or justifications of answers decreased performance levels.
- Performance levels in measurement increased significantly from 1990 to 1992 for students in all three grades. Within each year, performance increased steadily across grade levels, with a greater increase from grade 4 to grade 8 than from grade 8 to grade 12.

as scale, ruler, and thermometer. Two items were administered only to fourth-grade students, one item only to eighth-grade students, and one item only to twelfth-grade students. Students at all three grade levels were generally successful in choosing the correct measurement instrument from a list of choices, with percent-correct values ranging from 69 percent to 87 percent.

Six items asked students to read measurements from pictorial representations of instruments such as rulers, scales, thermometers, gauges, and protractors. None of these items required handling the instrument itself. Results for these six items are summarized in table 6.1 according to the measurement instrument pictured in each item. In general, eighth- and twelfth-grade students were successful in obtaining correct readings from pictures of measurement instruments. For example, 87 percent of the eighth-grade students correctly answered a regular constructed-response item that required reading a weight from a scale, and 91 percent of the twelfth-grade students selected the correct measure for an angle superimposed on a picture of a protractor. However, students in grade 4 had more difficulty reading measurement instruments, especially in two instances: (1) when the object to be measured was not positioned at the beginning of a ruler and (2) when the scale depicted on the instrument involved increments other than 1.

In the first instance, when fourth-grade students were asked on a secure, regular constructed-response item to write the correct measure of an object not aligned with the beginning of a ruler but with some other point, only about one-fourth of the students answered correctly. The most popular answer was that associated with the rightmost point of the object. For

Table 6.1 Performance on Items Involving Reading Measurement Instruments

Instrument Pictured in Item	Percent Correct		
	Grade 4	Grade 8	Grade 12
1. Thermometer	31	—	—
2. Scale	—	87	—
3. Gauge	—	57	—
4. Protractor	—	—	91
5. Scale	44	79	—
6. Ruler	24	62	82

Note: Items 1, 2, and 6 were regular constructed-response items, and the rest were multiple-choice items.

example, consider a small, sharpened pencil positioned parallel to a 6-inch ruler with the eraser end of the pencil at the ruler's 3-inch mark and the point at the 8-inch mark. Although the correct measure of the pencil would be 5 inches, a common error would be for students to answer 8 inches, the position of the pencil's point. This error pattern has been documented in research studies (Hart 1984) and reported for previous NAEP assessments (Lindquist and Kouba 1989). To help students in reading rulers in complex situations like the one just described, Reys, Suydam, and Lindquist (1995) suggest classroom activities involving direct comparison between the ruler and the object, such as having students mark off the units on the ruler corresponding to the object's length and then count them.

The multiple-choice item in table 6.2 illustrates the difficulty fourth-grade students had in reading measurement instruments with scales based on increments other than 1. The scale depicted on the measurement instrument represents an increment of 10 pounds for each short tick mark, with every 50 pounds labeled and indicated by a longer tick mark. Fewer than half the fourth-grade students chose the correct answer, with the majority

Table 6.2 Instrument with Scale in Increments Other than 1: Sample Item

| Item | Percent Responding | |
| | Grade 4 | Grade 8 |

What is the weight shown on the scale?

A. 6 pounds	2	1
B. 7 pounds	<1	<1
C. 51 pounds	53	20
D. 60 pounds*	44	79

*Indicates correct response.
Note: Percents may not add to 100 because of rounding or omissions.

choosing the distracter based on an increment of 1. This pattern recurs for a secure multiple-choice item that showed a thermometer with an increment of 2 degrees for each tick mark and asked fourth-grade students to select the correct temperature. On that thermometer item, 31 percent of the fourth-grade students selected the correct answer, but 50 percent chose the distracter based on an increment of 1. These results suggest that elementary school students may not be attending closely to the increments used on a scale; rather, they think almost exclusively in terms of an increment of 1. However, there is evidence from NAEP that by the middle grades, students appear to be attending more closely to scale on measurement instruments. In particular, for the item in table 6.2 that was also administered to eighth-grade students, 79 percent of those students chose the correct answer, with only 20 percent choosing the distracter based on an increment of 1. Given the NAEP findings on the scaling errors made by fourth-grade students, students in the elementary grades should be given additional experience in working with measurement instruments with scales based on a variety of increments.

In addition to items that required students to read measurement instruments, the 1992 NAEP mathematics assessment included five regular constructed-response items. Fourth-grade students were provided with rulers and eighth- and twelfth-grade students with combination ruler/protractors, and then were asked to use these instruments to determine linear or angular measure. In general, students in grades 4 and 8 were successful in using rulers to obtain linear measurements, but eighth-grade students had more difficulty obtaining correct angular measurements using a protractor. At grade 12, only about 30 percent of the students could obtain correct linear and angular measurements for circular figures.

The items in table 6.3 are presented as examples of fourth- and eighth-grade students' performance on items requiring the use of measurement instruments. For the first two items in the table, more than 50 percent of fourth-grade students and more than 70 percent of eighth-grade students obtained accurate measurements using a centimeter ruler. However, for item 3 only 35 percent of the eighth-grade students obtained the correct degree measure for the angle depicted. One possible explanation of the low performance level on that item involves the way in which the angle to be measured was oriented; that is, with the angle in the upper left corner ($\angle$) rather than in the lower left corner ($\angle$), the orientation commonly used in textbooks and curricular materials. Findings from research studies (e.g., Clements and Battista 1992) suggest that students often have preconceived notions about angles and other geometric figures that are based solely on how the figures appear in textbooks. To lessen any impact that the figure's orientation may have, students should have experience in measuring angles presented in a variety of orientations. However, the

Table 6.3 Using Measurement Instruments: Sample Items

Item	Percent Correct	
	Grade 4	Grade 8

Directions for Items 1 and 2 below:

Use your centimeter ruler to make the following measurements to the <u>nearest centimeter</u>.

1. What is the length in centimeters of one of the longer sides of the rectangle? — 52 — 71

2. What is the length in centimeters of the diagonal from *A* to *B*? — 60 — 79

3. Use your protractor to find the degree measure of the angle shown above. — — — 35

Note: Figures in the table have been reduced from their original sizes.

preceding discussion assumes that students in mathematics classrooms have had the opportunity to use protractors as measuring devices, but data from the NAEP student background questionnaire suggest that this assumption may not be correct. When students were asked how often they worked with measuring instruments such as protractors, 52 percent of the eighth-grade students reported *never* working with them. Thus, another plausible explanation for the poor performance on the items requiring the use of a protractor involves the opportunity to learn.

SELECTING AND CONVERTING BETWEEN UNITS OF MEASUREMENT

Four multiple-choice items—two items administered only to fourth-grade students, one item only to twelfth-grade students, and one item to both fourth- and eighth-grade students—assessed the ability to select appropriate units of measurement in real-world situations. These are some examples: "Which unit would you use to measure the length of a pencil—centimeter, meter, or kilometer?" and "The length of a trail that Pat hiked in one day could have been ... 5 millimeters, 5 centimeters, 5 meters, or 5 kilometers?" Students in all three grades have a sense of which measurement unit would be appropriate for a particular situation, with percent-correct values ranging from 59 percent to 88 percent. Students did well not only on items involving customary units but also on those involving metric units.

Five multiple-choice items measured students' ability to convert between units of measure within the same system. One item was administered only to fourth-grade students, two items only to twelfth-grade students, one item to both fourth- and eighth-grade students, and one item to both eighth- and twelfth-grade students. For all items, the conversions were restricted to customary units of linear measure, liquid capacity, or time, and conversion factors such as 3 feet = 1 yard and 1 gallon = 4 quarts were given. Patterns in responses suggest that students in grades 8 and 12 can perform simple, one-step conversions but that students in all three grades have difficulty solving more-complex conversion problems. For two secure items dealing with units of time in a real-world context, percent-correct values for fourth-grade students ranged from 23 percent to 36 percent. These results suggest that these students have difficulty converting the passage of hours and fractions of hours into actual time; for example, knowing that if a person began walking at 8:30 a.m., walked for 4 1/2 hours, and then stopped, the time at which the person stopped would be 1:00 p.m. On items involving linear measures, eighth- and twelfth-grade students can do simple conversions, with percent-correct values ranging from 84 percent at grade 8 to 91 percent at grade 12. However, only about half the twelfth-grade students could successfully answer an item involving more than one conversion.

PERIMETER, AREA, VOLUME, AND SURFACE AREA

The 1992 NAEP mathematics assessment included twenty-eight items that assessed the topics of perimeter, area, volume, and surface area. Performance on each of these topics is discussed in the sections that follow.

Perimeter

Table 6.4 contains descriptions of the six perimeter items on NAEP and their corresponding percent-correct values. Patterns in fourth- and eighth-grade students' responses suggest that as perimeter problems become more complex, students' performance levels decrease. For example, many students in grade 4 could determine the amount of fencing needed to "go around" a rectangular garden, and most students in grade 8 could determine the length of one side of a geometric figure given its perimeter, but students in those grades had difficulty with perimeter items involving comparisons between two or more geometric figures. For both eighth- and twelfth-grade students, requiring written explanations or justifications of answers to constructed-response questions also added to the difficulty levels of the items and decreased performance. For example, item 5, a regular constructed-response item described in table 6.4, required students to choose the figure with the longest perimeter and then explain their choice. Although a majority of the eighth- and twelfth-grade students could identify correctly the figure with the longest perimeter, only a small percentage could justify or explain why their choice was correct.

Table 6.4   Performance on Perimeter Items

	Percent Correct		
Item Description	Grade 4	Grade 8	Grade 12
1. Choose the correct value for the amount of fencing needed to go around a rectangular garden.	46	—	—
2. Use information about the perimeters of geometric figures to choose the correct value for the length of a side of one of the figures.	20	—	—
3. Choose which geometric figure could have a perimeter of 28 units.	—	32	—
4. Estimate the perimeter of a geometric figure.	—	19	—
5. Determine which geometric figure has the longest perimeter and explain the answer.	—	5	13
6. Choose the correct length of one side of a geometric figure, given its perimeter.	37	69	76

Note: Items 4 and 5 were regular constructed-response items, and the rest were multiple-choice items.

Further, performance patterns on the perimeter items suggest that understanding of perimeter concepts increases more sharply between grades 4 and 8 than between grades 8 and 12. For example, for item 6 in table 6.4 there is a 32 percentage-point difference in performance between fourth- and eighth-grade students, whereas the performance difference between eighth- and twelfth-grade students is only 7 percentage points. This pattern of performance differences between contiguous grade levels in NAEP seems to be in alignment with the usual instructional sequence of introducing perimeter in grade 3 or grade 4 and continuing its development through grade 8.

The two multiple-choice items in table 6.5 illustrate some difficulties that fourth- and eighth-grade students have with perimeter concepts. Performance by fourth-grade students on item 1 in the table is both encouraging and troublesome. The encouraging part is that almost half the fourth-grade students have some understanding of the concept of perimeter for rectangles in a real-world context, but the troublesome aspect is that over half of the fourth-grade students appear not to have even a basic understanding of perimeter. A careful look at the response choices provides some insight into students' misunderstandings and error patterns. Choice A, selected by more than one-third of the students, involves adding the two numbers that appear in the diagram and obtaining an answer of 18 feet. One could conjecture that some students may have used an incorrect, fallback strategy of "when in doubt, add." Students either solved the problem by working with the numbers supplied in the figure without reading the problem or did not understand the problem when they read it and "fell back" to an inappropriate addition strategy. Students also may have realized that "go around" implies the concept of perimeter but not that they must use the lengths of the unlabeled sides in determining the correct answer. Giving this item—either in its present multiple-choice form or as a constructed-response question—to elementary school students and then discussing their solution strategies could serve as a way to explore students' approaches to solving simple perimeter problems.

Students also had difficulty working with perimeter in an abstract geometric context, as illustrated by item 2 in table 6.5. Only about one-third of the eighth-grade students correctly identified figure B as the figure that could have a perimeter of 28 units. This item requires an understanding of perimeter and geometric relationships or practiced skill in visual estimation. For example, realizing that figure B could have a perimeter of 28 units involves knowledge that opposite sides of rectangles have equal lengths and possibly an understanding of the Pythagorean relationship for special triangles (here, a 3-4-5 triangle). To identify choices A, C, and D as incorrect requires that students use what they know about the equality of opposite sides of rectangles to deduce that figure A and figure D could have

Table 6.5 Perimeter Items

Item	Percent Responding	
	Grade 4	Grade 8

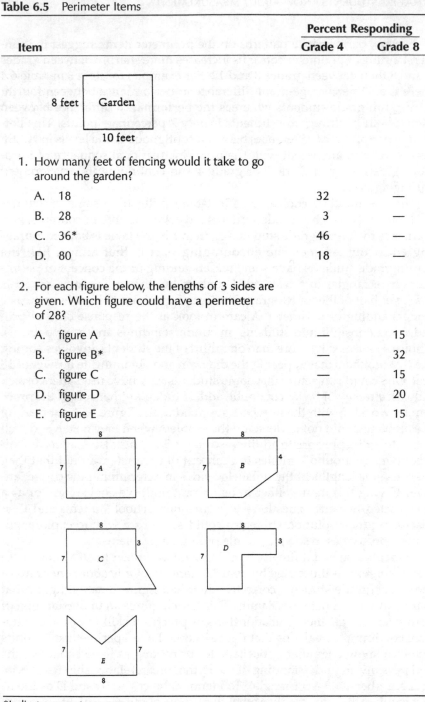

1. How many feet of fencing would it take to go around the garden?

A. 18	32	—
B. 28	3	—
C. 36*	46	—
D. 80	18	—

2. For each figure below, the lengths of 3 sides are given. Which figure could have a perimeter of 28?

A. figure A	—	15
B. figure B*	—	32
C. figure C	—	15
D. figure D	—	20
E. figure E	—	15

*Indicates correct response.

Note: Percents may not add to 100 because of rounding or omissions.

perimeters of 30 units and that figure C could have a perimeter of more than 30 units. To identify choice E as incorrect requires the recognition that the sum of the lengths of any two legs of a triangle is always greater than the length of the third side (the triangle inequality). It might be interesting to have middle school students produce solutions to this problem and then interview these students to assess whether they are just "seeing" these relationships without being able to articulate them or whether students are cognizant of the geometric relationships and their effect on perimeter and also can describe those relationships in either formal or informal terms.

Area

Of the sixteen area items on NAEP, nine were administered at more than one grade level, thus enabling the identification of performance patterns across grades. Results for these nine cross-grade items appear in table 6.6, with some items involving simple area concepts (for example, items 1 and 2) or the use of the formula for finding area (for example, item 5), and other items assessing area in more complex situations. About two-fifths to three-fourths of the fourth-grade students and two-thirds to three-fourths of the eighth-grade students have an understanding of simple area concepts, but only one-fifth of the fourth-grade students were successful in using the area formula to solve a problem. Patterns in performance on the set of items in table 6.6 and on the other area items suggest that many students have developed incomplete conceptual understanding of area and appropriate area units and have difficulty in applying area to complex situations.

Understanding Area Units

An important concept for students working with area is the nature of the appropriate units of measure. Students with a deep understanding of area know that area units are expressed as square units, they "take up a flat space," and they can be decomposed and recombined to determine areas of nonrectangular figures. Further, when arranged in arrays, area units can be easily counted by using length measurements and multiplication. In fact, the formula "length × width = area" is the short form of a longer conceptual definition: The number of square units in a row times the number of rows is equal to the total number of square units (Reys, Suydam, and Lindquist 1995).

NAEP results indicate that some fourth-grade students and most eighth-grade students understand that area is measured in square units. In particular, responses to a secure multiple-choice item, described in table 6.6 as item 3, requiring students to find the area of a polygon depicted on a grid show that about two-fifths of the fourth-grade students and about three-fourths of the eighth-grade students were able to conceive of area in terms

Table 6.6 Performance on Area Items

	Percent Correct		
Item Description	**Grade 4**	**Grade 8**	**Grade 12**
1. Choose the polygon that has the least area.	72	74	—
2. On a grid, draw a rectangle with an area of 12 square units.	42	66	—
3. Choose the correct value for the area of a polygon depicted on a grid.	39	73	—
4. Choose the correct value for the area of a rectangular carpet, when the length and width are given.	19	65	—
5. Choose the correct algebraic representation for area.	—	37	52
6. Choose the correct value for the amount of carpeting needed to cover a portion of a rectangular room.	—	29	52
7. Choose the correct value for the area between a square and a circle inscribed in the square.	—	29	37
8. Determine which geometric figure has the greatest area and explain the answer.	—	23	32
9. Determine the area of a trapezoid.	—	9	23

Note: Items 2, 8, and 9 were regular constructed-response items, and the rest were multiple-choice items.

of square units that could be decomposed and recombined. Further evidence of understanding of area as measured in square units can be found in the results for item 1 in table 6.7. When given a grid, about two-fifths of the fourth-grade students and two-thirds of the eighth-grade students successfully drew a rectangle with an area of 12 square units. Very few students drew a rectangle with a perimeter of 12 units. Looking at a set of student responses to this constructed-response item revealed that the fourth-grade students tended to draw 1-by-12 rectangles and that, although some eighth-grade students also drew 1-by-12 rectangles, those students produced more 6-by-2 rectangles or 4-by-3 rectangles than students in grade 4. The question arises, then, whether students who drew 1-by-12 rectangles understood that they were outlining a rectangle of 12 square units or whether they were just outlining 12 squares.

Table 6.7 Understanding Area Units: Sample Items

Item	Percent Responding	
	Grade 4	**Grade 8**

1. On the grid below, draw a rectangle with an area of 12 square units.

☐ = 1 square unit

	Grade 4	Grade 8
Any rectangle with an area of 12 square units*	42	66
Any rectangle with a perimeter of 12 units	2	1
Any other incorrect response	48	30
Omitted	7	3

2. A rectangular carpet is 9 feet long and 6 feet wide. What is the area of the carpet in square feet?

	Grade 4	Grade 8
A. 15	45	16
B. 27	16	5
C. 30	19	13
D. 54*	19	65

*Indicates correct response.

Note: The figure in item 1 has been reduced from its original size. Percents may not add to 100 because of rounding or omissions.

The results for item 2 in table 6.7 suggest that using multiplication to find area in a real-world context was difficult for students in grade 4. Only 19 percent of the fourth-grade students selected the correct answer; students in grade 8 performed much better, with 65 percent choosing the correct answer. For fourth-grade students, the most attractive distracter was choice A, which is obtained by adding the numbers given in the problem and may be taken as further evidence that many students in grade 4 use addition strategies inappropriately. About one-fifth of the fourth- and eighth-grade students selected choice C, representing the perimeter of the figure. These results might be attributed to guessing, but they might also suggest that some students were confusing area and perimeter. This finding stands in contrast to the result reported for item 1, in which *few* students confused area and perimeter. One conjecture about these contrasting results is that the use of the grid for the constructed-response item may have cued students to focus on area rather than perimeter. Other differences in the structures of the two items, such as the fact that only one number (12) appears in item 1 whereas item 2 has two numbers (6 and 9) and that one item is multiple-choice and the other constructed-response, may also have influenced performance levels.

Performance on a secure multiple-choice item requiring the selection of the correct numerical representation for solving an area problem, described in table 6.6 as item 5, suggests that eighth- and twelfth-grade students have difficulty symbolically representing the calculation of area as the multiplication of linear measurements. Only 37 percent of the eighth-grade students and 52 percent of the twelfth-grade students chose the correct symbolic representation. Further, students in these two grades continue to confuse area and perimeter. On the secure item just described, at both grade levels the most popular incorrect choice was the symbolic representation for perimeter, selected by 37 percent of the eighth-grade students and 19 percent of the twelfth-grade students.

Applying Area Concepts

The three released items shown in table 6.8 required eighth- and twelfth-grade students to apply area concepts in real-world and abstract mathematical contexts. These items were in blocks for which students were permitted to use scientific calculators provided by NAEP. Across the three items, students in grade 8 had difficulty with applied area problems. On the two multiple-choice items, eighth-grade students performed at a level just above that expected from guessing, and on the constructed-response item, their performance was very low. The patterns for students in grade 12 show that they were more successful in solving complex problems based on finding the areas of rectangles (item 1) than in solving problems based on finding the area of two geometric figures, one of which was nonrectangular (items 2 and 3). As with the eighth-grade students, the twelfth-grade students also had a lower performance level on the constructed-response item than on the multiple-choice items.

Performance results for item 1 show that almost one-third of the eighth-grade students and slightly more than half the twelfth-grade students chose the correct answer. However, a closer look at the dimensions of the room to be carpeted and the intermediate calculations involved in obtaining the correct answer reveals a possible relationship between the dimensions, the intermediate calculations, and the correct response that involves the digit 8. In particular, the width of the room is 8.5 feet; the length of one side of the carpeted area is 10.5 feet – 2.5 feet = 8 feet; and the correct answer is about 8 square yards (emphases added). Because of the recurrence of the digit 8, then, it is impossible to determine whether students arrived at the correct answer by subtracting two numbers given in the problem, selecting the answer closest in value to the given width of the room, or solving the problem completely.

Item 2 in table 6.8 is representative of the standard inscribed-circle, "find the shaded area" problems found in most textbooks. Yet, despite the possibility that students may have seen this kind of problem before in mathematics class, fewer than one-third of the eighth-grade students and fewer than two-fifths of the twelfth-grade students answered this item correctly. For students at both grade levels, the most popular distracter was choice C, the area of the circle. However, because of the item's multiple-choice format, it is impossible to know whether students chose this answer by merely guessing or if they were completing only one part of a multistep problem.

This item was classified by NAEP as being "calculator-active" (that is, the solution required the use of a calculator), and performance data were obtained that compared the percent-correct values for students based on their yes or no responses to the self-report question "Did you use the calculator on this question?" At grade 8, the percent-correct value for students who said "yes" to the self-report question was 37 percent and the corresponding value for students who said "no" was 28 percent, with the difference between the values statistically significant at the .05 level. At grade 12, the percent-correct values for students who responded affirmatively and those who responded negatively were 53 percent and 31 percent, respectively, with the difference also statistically significant at the .05 level. For both grades, then, students who reported using the calculator did better on the item than students who reported not using the calculator. These results reflect a general trend reported by Hembree and Dessart (1986), that is, the use of calculators in testing produces higher achievement scores than the use of paper-and-pencil calculations. However, the results associated with the NAEP item just described must be interpreted with caution because of a variety of factors, including the guessing factor inherent in

Table 6.8 Applying Area Concepts: Sample Items

Item[a]	Percent Responding	
	Grade 8	Grade 12

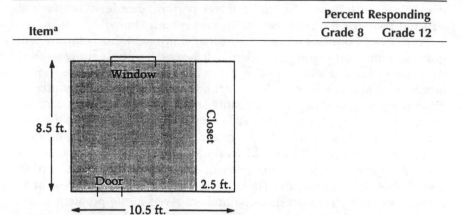

1. Chris wishes to carpet the rectangular room shown above. To the nearest <u>square yard</u>, how many square yards of carpet are needed to carpet the floor of the room if the closet floor will not be carpeted?

   (1 square yard = 9 square feet)

	Grade 8	Grade 12
A. 8*	29	52
B. 10	18	16
C. 11	10	6
D. 19	22	10
E. 22	19	13

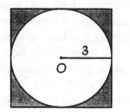

2. In the figure above, a circle with center O and radius of length 3 is inscribed in a square. What is the area of the shaded region?

	Grade 8	Grade 12
A. 3.86	21	11
B. 7.73*	29	37
C. 28.27	22	24
D. 32.86	8	7
E. 36.00	13	11

Item[a]	Percent Responding	
	Grade 8	Grade 12

3. The area of rectangle BCDE shown above is 60 square inches. If the length of AE is 10 inches and the length of ED is 15 inches, what is the area of trapezoid ABCD, in square inches?

	Grade 8	Grade 12
Correct response of 80 square inches	9	23
Any incorrect response	81	67
Omitted	9	10

*Indicates correct response.
[a]Items are from item blocks for which students were provided with, and permitted to use, scientific calculators.
Note: Percents may not add to 100 because of rounding or omissions.

multiple-choice items and the lack of direct information about how students used the calculator on this item. One possible conjecture about the performance differences is that because of the need for a numerical approximation for π, which could be obtained automatically from a key on the scientific calculator, using the calculator may have increased accuracy in computing the area of the circle and thus affected the degree to which the computed answer more closely matched the correct answer choice.

As noted previously, performance levels on item 3 in table 6.8 were very low for both eighth- and twelfth-grade students. Because students' work for that item was not analyzed here, it is impossible to identify with certainty the strategies students used to solve this problem or the reasons why they had difficulty arriving at a correct answer. The low performance levels associated with item 3 suggest possible issues to investigate. One issue involves the way in which the item was worded. In their article about designing open-ended mathematics tasks for middle school mathematics classrooms, Santel-Parke and Cai (forthcoming) report that a task's wording can influence the strategies students use and the quality of their responses. In item 3, then, did the use of the word *trapezoid* lead students to believe that the only way to solve the problem was through the use of the formula for finding the area of a trapezoid? What if the question were phrased in a more open fashion, that is, "What is the area of figure ABCD?" It might be interesting to give both versions of this item to middle school students or high school students and then analyze their results to see whether the wording had any influence on the choice of strategy.

Volume and Surface Area

Of the three items that assessed volume, one was administered only to fourth-grade students and two items only to twelfth-grade students. There were no volume-related items on the eighth-grade test. One surface-area item was administered to eighth-grade and twelfth-grade students, and two complex items assessing both volume and surface area were administered only to twelfth-grade students. Because of the dearth of items given to students in grade 4 and grade 8, little can be concluded about those students' understanding of, or difficulties with, volume or surface area. For students in grade 12, results on items assessing students' understanding of simple volume concepts and their ability to calculate volume when given the appropriate formula ranged from 53 percent to 68 percent correct. However, complex volume items and items assessing surface area concepts were difficult for twelfth-grade students.

On item 1 in table 6.9 more than half the twelfth-grade students had little difficulty working with the concept of volume, but the results also suggest that many students continued to have difficulty attending to more than one condition in a problem. Almost one-third of the students selected choice B (512), which represents only *half* the volume of the box. Those students may either have found the volume of half the box but did not double it or have not realized that the number of cubes given in the problem represented only half the number of cubes needed to fill the box.

Only about 30 percent of the twelfth-grade students responded correctly to item 2, a complex item that required either recalling and applying the formula for the volume of a cylinder or understanding that the radius is the most influential attribute in determining the order of the cylinders by volume. That 44 percent of the students selected choice A ($y = w = x$) suggests that those students probably solved the item by multiplying the height and the radius for each cylinder. This could reflect either the "when in doubt, multiply" strategy or an indication of some confusion between the concepts of area and volume.

Surface area was a very difficult concept for the twelfth-grade students. On the two secure constructed-response items involving surface area, fewer than 10 percent of the twelfth-grade students gave correct answers. Thus, little insight on students' understandings and misunderstandings of surface area can be gleaned from NAEP results.

OVERALL PERFORMANCE IN MEASUREMENT

This chapter has discussed student performance on individual measurement items and clusters of related items. Gaining a more global view of student performance for measurement as a NAEP content area requires looking at performance across the entire set of NAEP measurement items.

Table 6.9 Volume Items

Item	Percent Responding Grade 12
1. It takes 64 identical cubes to <u>half</u> fill a rectangular box. If each cube has a volume of 8 cubic centimeters, what is the volume of the box in cubic centimeters?	
A. 1,024*	53
B. 512	30
C. 128	6
D. 16	4
E. 8	5

Volume = w Volume = x Volume = y

Item	Percent Responding Grade 12
2. In the figures above, the radius and height of each right circular cylinder are given. If w, x, and y represent respective volumes of the cylinders, which of the following statements is true?	
A. $y = w = x$	44
B. $y < x < w$*	30
C. $y < w < x$	7
D. $w < y < x$	3
E. $w < x < y$	15

*Indicates correct response.
Note: Percents may not add to 100 because of rounding or omissions.

In NAEP, overall performance for measurement is reported as an average performance score on a scale from 0 to 500. Table 6.10 presents data on the average performance in measurement for students in grades 4, 8, and 12 for 1990 and 1992 and includes data for selected subgroups.

Overall performance in measurement significantly increased from 1990 to 1992. In both years, performance in measurement increased steadily from grade 4 to grade 12, with a greater increase from grade 4 to grade 8 than from grade 8 to grade 12. One possible explanation for the greater performance increases from the elementary grades to the middle grades in 1992 can be found in the teachers' responses to a particular question about their instructional practices regarding measurement topics, "How much emphasis did you or will you give to the topic of measurement in your mathematics class?" In 1992, about 94 percent of the fourth-grade students had teachers who reported giving either heavy or moderate emphasis to measurement. However, for eighth-grade students, the percent of their teachers who reported giving heavy or moderate emphasis to measurement concepts dropped 10 percentage points to 84 percent. Focusing on measurement in the upper elementary grades, then, appears to foster student understanding that carries over into the middle grades and results in greater increases in NAEP measurement scores from grade 4 to grade 8. However, as emphasis on measurement decreases in the middle grades, performance gains on measurement in NAEP from grade 8 to grade 12 have a lower rate of increase.

For males and females in both 1990 and 1992, performance levels in measurement were nearly identical for all three grade levels, and those levels were also similar to the overall performance levels for each year. Females in all three grade levels performed at a significantly higher level in 1992 than in 1990; males in grades 4 and 8 performed better in 1992 than in 1990. For the race/ethnicity subgroups, White students' performance in measurement increased from 1990 to 1992 in grades 4, 8, and 12. For Black students the increase in performance was significant only at grade 12. The 1992 NAEP results reveal disparities in performance in measurement between race/ethnicity subgroups: White students scored significantly higher than both Black and Hispanic students, and Hispanic students scored significantly higher than Black students.

CONCLUSION

The NCTM *Curriculum and Evaluation Standards for School Mathematics* (1989) includes standards for measurement as a content area for grades K–4 (pp. 51–52) and grades 5–8 (pp. 118–19), with the implicit assumption that measurement concepts learned in the earlier grades appear throughout the high school curriculum. The *Standards* and the NAEP framework

Table 6.10 Average Performance in Measurement: Overall Performance and Performance by Gender and Race/Ethnicity, Grades 4, 8, and 12 for 1990 and 1992

	Grade 4	Grade 8	Grade 12
Overall			
1992	224 (0.8)>	266 (1.2)>	297 (0.9)>
1990	218 (1.0)	259 (1.6)	292 (1.3)
Gender			
Male			
1992	225 (1.0)>	268 (1.4)>	301 (1.2)
1990	221 (1.3)	263 (2.0)	298 (1.5)
Female			
1992	222 (1.1)>	264 (1.5)>	294 (1.1)>
1990	216 (1.3)	255 (1.5)	288 (1.5)
Race/Ethnicity			
White			
1992	233 (1.0)>	277 (1.3)>	305 (1.0)>
1990	227 (1.3)	268 (1.7)	299 (1.4)
Black			
1992	194 (1.8)	226 (1.9)	268 (1.9)>
1990	190 (2.3)	228 (3.2)	262 (2.2)
Hispanic			
1992	204 (1.7)	243 (1.8)	281 (1.8)
1990	204 (2.3)	238 (3.3)	276 (3.0)

> The value for 1992 was significantly higher than the value for 1990 at about the 95 percent confidence level. The standard errors of the estimated proficiencies appear in parentheses.

emphasize many of the same measurement concepts, such as measurement attributes, units of measurement, the real-world applications of measurement, the importance of being able to communicate about measurement results, and the interconnectedness of measurement and geometry. In light of these connections between the NCTM *Standards* and the NAEP framework, results from NAEP can be viewed as one indicator of what students know and can do with respect to measurement concepts and skills identified in the NCTM *Standards*.

Results from the 1992 NAEP mathematics assessment reveal that students at all three grade levels were able to read measurement instruments, use simple measurement instruments, and select appropriate units of measurement, all of which are emphasized in the NCTM *Curriculum and Evaluation Standards*. However, students had more difficulty demonstrating their understanding of important concepts in perimeter, area, volume, and surface area, particularly in complex situations involving real-world applications and multistep problems. Performance was especially low when students had to provide an explanation of their answer. Overall results show growth in achievement in measurement from 1990 to 1992, but there is also evidence of differential performance in measurement between minority students and majority students. These results suggest that in order to satisfy more completely the expectations in the NCTM Curriculum Standards for measurement, *all* students should have more experience in working with measurement concepts in complex, problem-solving situations and in communicating about measurement results.

REFERENCES

Clements, Douglas H., and Michael T. Battista. "Geometry and Spatial Reasoning." In *Handbook of Research on Mathematics Teaching and Learning,* edited by Douglas A. Grouws, pp. 420–64. New York: Macmillan, 1992.

Hart, Kathleen M. "Measurement." In *Children's Understanding of Mathematics: 11–16,* edited by Kathleen M. Hart, pp. 9–22. London: John Murray Publishers, 1984.

Hembree, Ray, and Donald J. Dessart. "Effects of Hand-Held Calculators in Precollege Mathematics Education: A Meta-analysis." *Journal for Research in Mathematics Education* 17 (March 1986): 83–99.

Lindquist, Mary M., and Vicky L. Kouba. "Measurement." In *Results from the Fourth Mathematics Assessment of the National Assessment of Educational Progress,* edited by Mary M. Lindquist, pp. 35–43. Reston, Va.: National Council of Teachers of Mathematics, 1989.

National Assessment of Educational Progress. *Mathematics Objectives: 1990 Assessment.* Princeton, N.J.: Educational Testing Service, National Assessment of Educational Progress, 1988.

National Council of Teachers of Mathematics. *Curriculum and Evaluation Standards for School Mathematics.* Reston, Va.: National Council of Teachers of Mathematics, 1989.

Reys, Robert E., Marilyn N. Suydam, and Mary M. Lindquist. *Helping Children Learn Mathematics*. 4th ed. Boston: Allyn & Bacon, 1995.

Santel-Parke, Carol, and Jinfa Cai. "Does the Task Truly Assess What Was Intended?" *Mathematics Teaching in the Middle School* (forthcoming).

7

What Do Students Know about Geometry?

Marilyn E. Strutchens & Glendon W. Blume

GEOMETRIC SHAPES, ideas, and concepts are experienced by students daily in the real world and thus are important components of the kindergarten through grade 12 curriculum. Geometric ideas not only span the mathematics curriculum but are connected to other subjects as well. This chapter discusses what students know and are able to do in the content area of geometry according to NAEP results. The first section describes geometry in NAEP and includes information about the set of geometry items on the 1992 assessment. The next sections give details about student performance on items that assessed specific topics in geometry, and the final sections present information on overall performance on geometry and NAEP and some conclusions that can be made from NAEP results.

GEOMETRY IN NAEP AND THE NAEP GEOMETRY ITEMS

For the content area of geometry in the 1992 assessment, NAEP focused on "students' knowledge of geometric figures and relationships and on their skills in working with this knowledge" (National Assessment of Educational Progress 1988, p. 23). In agreement with the *Curriculum and Evaluation Standards for School Mathematics* produced by the National Council of teachers of Mathematics (1989), NAEP's geometry items tested students' ability to model and visualize geometric figures in one, two, and three dimensions and to communicate geometric ideas. In addition, students' ability to use informal reasoning and to establish geometric relationships, which are also skills advocated by the *Standards,* were tested.

The 1992 assessment contained sixty-seven items related to geometry. A subset of the geometry items was administered at more than one grade level or appeared on both the 1990 and 1992 mathematics assessments, which facilitated comparisons of students' performance between grade levels and changes in performance between 1990 and 1992. For some items, students in grade 4 were provided with paper rulers and students in grades

8 and 12 were provided with combination ruler/protractors. To answer other items, students were given cardboard manipulatives in the form of geometric shapes. The following sections describe performance on these specific topics and skills related to geometry: drawing, constructing, and visualizing geometric figures; identification of geometric figures; and analysis and application of geometric figures and properties.

DRAW, CONSTRUCT, AND VISUALIZE GEOMETRIC FIGURES

Drawing, constructing, and visualizing geometric figures are skills that help students to discover mathematical relationships and develop spatial sense (NCTM 1989). As suggested by the *Standards*, in an appropriate learning environment, students should be able to make conjectures and derive definitions from experiences with drawing, constructing, and visualizing two- and three-dimensional figures. A subset of the NAEP geometry items fell into two categories: (1) drawing and constructing and (2) visualization. In the first category students were asked to draw figures using manipulatives, to outline figures embedded in a large shape made of triangles, to draw figures given specific numbers of sides and types of angles, and to use measuring tools to construct specific figures. The second category included items asking students to visualize what nets (flat shapes

HIGHLIGHTS

- As the geometric figures and the situations in the items became more complex, fourth-, eighth-, and twelfth-grade students' performance levels decreased.
- Based on the results from NAEP items, most of the students at all three grade levels appeared to be performing at the "holistic" level of the van Hiele levels of geometric thought.
- Students in grade 4 were able to identify properties of simple geometric figures but had more difficulty working with more-complex figures and providing more extensive written explanations.
- On items involving angle sums in triangles, performance of twelfth-grade students was considerably better than that of eighth-grade students.
- On items involving applications of geometric properties of circles or squares, students appeared to be more successful when familiar content or procedures were involved.
- Students in grade 12 have difficulty understanding some concepts associated with analytic geometry.
- Performance levels in geometry increased significantly from 1990 to 1992 for students in all three grades. Within each year, performance increased steadily across grade levels.

that will fold up to make a solid) would look like if the nets were folded into the solid, to determine if an object is symmetrical, to look at a solid figure and determine what its net would look like, to determine the results of various cuts of solids into solids or two-dimensional figures, and to determine the results of certain transformations. Performance results on the NAEP items involving drawing or constructing geometric figures and visualizing geometric figures are described in the next sections.

Drawing or Constructing Geometric Figures

There were fourteen items on NAEP that required students to draw or construct geometric figures. The majority were regular constructed-response items administered either at grade 4 only or at grades 4 and 8, and they assessed students' ability to draw simple geometric figures such as squares, rectangles, and triangles. Of the remaining items that assessed more-complex concepts such as drawing quadrilaterals other than rectangles and drawing perpendicular lines, three were administered only to eighth-grade students and two only to twelfth-grade students. Because only two items were administered to students in grade 12 and because performance was only 20 percent correct on each item, little can be discovered about what twelfth-grade students know about drawing or constructing geometric figures.

Table 7.1 contains representative performance levels for a subset of regular constructed-response items administered to students in the lower two grades. As expected, performance levels varied by grade level and according to the complexity of the drawing or construction required. For example, 65 percent of fourth-grade students could form a simple quadrilateral based on manipulative geometric shapes and about 40 percent could draw a square, given particular materials or conditions in the items. However, as the figures to be drawn became more complex (for example, triangles and nonrectangles such as parallelograms), fourth-grade students' performance was much lower, ranging from 35 percent to less than 10 percent correct. Most eighth-grade students had little difficulty measuring angles and drawing simple figures such as squares, rectangles, and parallelograms, with percent-correct values ranging from 84 percent to just over 50 percent. Eighth-grade students had the most difficulty with a multiple-choice item about a geometric construction; only about one-fourth of the students selected the correct answer.

A closer look at a few items can provide additional information on performance results for drawing and constructing geometric figures. Two released items, both administered to students in grades 4 and 8, asked students to draw quadrilaterals, a rectangle 2 inches wide by 3 1/2 inches long in one item and a square based on two given corner points in the other item.

Table 7.1 Performance on Items Requiring Drawing or Constructing Geometric Figures

	Percent Correct	
Item Description	Grade 4	Grade 8
1. Use manipulatives to form a simple quadrilateral.	65	84
2. Use manipulatives to form a given polygon.	52	78
3. Use a ruler to draw a square given two corner points.	40	67
4. Identify triangles in a figure.	35	—
5. Sketch a polygon with a given number of right angles.	29	—
6. Use a ruler to draw a rectangle with given dimensions.	18	58
7. Sketch an angle with given degree measure.	10	70
8. Identify a nonrectangle in a figure.	6	—
9. Show how a simple quadrilateral can be formed from given geometric shapes.	—	57
10. Show how a quadrilateral can be formed from given geometric shapes.	—	51

Note: All were regular constructed-response items.

While working on these items, students were provided with rulers or combination ruler/protractors. For the first item, 18 percent of the fourth-grade students and 58 percent of the eighth-grade students drew the rectangle correctly within a tolerance of ± 1/2 inch. Another 32 percent of the fourth-grade students and another 24 percent of the eighth-grade students answering this question were able to draw a rectangle, but with incorrect dimensions. Because student work for that item was not analyzed, it is impossible to identify with certainty the reasons why those students did not draw a rectangle with specified dimensions. These results may be related to some students' inability to use rulers accurately or to students' choosing to estimate the lengths of the sides instead of using the ruler provided.

Additionally, it may be interesting to compare the results on this construction item with two items classified as measurement by NAEP (see chapter 6). When asked to use rulers to find linear measures for a given rectangle, more than half the students in grade 4 and three-fourths of the students in grade 8 obtained the correct measurements. However, as shown by the performance results for the geometry item described above, it was much more difficult to construct a rectangle of given dimensions than to measure an already existing rectangle. Students' responses on these two items may indicate that students are asked to measure already existing shapes more often than they are asked to construct shapes with particular dimensions.

Performance on the other regular constructed-response item, hereafter referred to as Draw a Square and shown in table 7.2, was somewhat higher. Asked to draw a square given two corner points, 40 percent of the fourth-grade students and 67 percent of the eighth-grade students produced

Table 7.2 Draw a Square

	Percent Responding	
Item	**Grade 4**	**Grade 8**
In the space below, use your ruler to draw a square with two of its corners at the points shown.		
Correct answer: a square with the two dots as adjacent vertices	10	18
Correct answer: a square with the two dots as diagonal vertices	30	49
Any incorrect response	52	31
Omitted	7	2

Note: Percents may not add to 100 because of rounding.

correct figures—either squares with the given dots as adjacent vertices or as diagonal vertices. From NAEP results for both grade 4 and grade 8, it appears that more students used the given dots as diagonal vertices than as adjacent vertices. However, those results provide no information on the kinds of incorrect responses students gave to this question. This lack of information is particularly distressing at grade 4 given that over half the students gave incorrect responses.

To gain a better understanding of how fourth-grade students were thinking about the Draw a Square item, a sample representing about 16 percent of the total number of nonblank responses was gathered and analyzed for this chapter. The responses in the sample were evaluated for correctness by using the same techniques used by NAEP, that is, by using a template with a ± 1/16-inch tolerance, with the result that responses in the sample were about equally divided between correct and incorrect. As was the case for the overall results, more correct responses had the given points used as diagonal vertices than as adjacent vertices, but in the case of the incorrect responses, this situation was reversed; that is, more incorrect responses showed evidence of the given points used as adjacent vertices. Responses 1 and 2 in figure 7.1 are examples of incorrectly drawn, squarelike figures using the points as adjacent vertices of the square. This pattern of responses in the sample gathered for this chapter suggests that using the points as diagonal points led to more correct answers than viewing the points as adjacent, a finding that was confirmed by a chi-square test that was significant at the .05 level. In thinking about why the diagonal-point view was more successful than the adjacent-point view, it could be that seeing the points as diagonals facilitated drawing the square because in this case the sides of the square were basically parallel to the pages of the test booklet. Thus, there is evidence that performance on the Draw a Square item was influenced by the way in which students were visualizing the square in relation to the given dots.

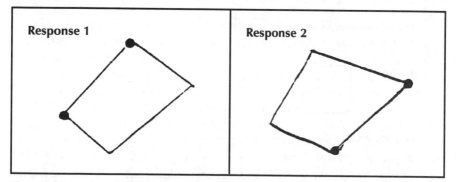

Fig. 7.1 Examples of responses to Draw a Square

A multiple-choice item administered to eighth-grade students assessed their understanding of a procedure involving a construction: given a horizontal line, students were asked to select the correct instructions for constructing a 45° angle at a particular point on the line. No actual constructions were needed to answer the question. Only 24 percent of the students chose the correct response, while 28 percent chose the directions for constructing a right angle. On the basis of the performance on this item and given that it was the only item of its kind on the 1992 NAEP, one cannot say that students really understood how to construct a 45° angle with the given conditions or if students just guessed.

Results from the set of NAEP items just discussed suggest that students in the elementary and middle grades need more experience with constructing geometric figures. Dynamic geometric construction software such as the Geometer's Sketchpad can be used in classrooms to provide students with the experiences needed for developing an intuitive and inductive understanding of geometric relations (Olive 1991). The Geometer's Sketchpad and similar programs provide students with an opportunity to experiment, observe, record, and conjecture with animated data, which enables students to develop a better understanding of geometric figures and their constructions.

Visualizing Geometric Figures

Ten NAEP items assessed students' ability to visualize in geometric situations. Two items were administered only to students in grade 4, two only to students in grade 12, and the rest to students across two or more grade levels. At the two lower grades, most students could visualize about simple geometric situations, such as those involving symmetry, unfolding a piece of paper, and flipping an object to another position. Here, percent-correct values were 65 percent for fourth-grade students and 87 percent for eighth-grade students. However, at all three grade levels students had more difficulty visualizing more-complex geometric figures, especially those presented in three dimensions.

Several NAEP items involving visualization appear in table 7.3. On item 1, 65 percent of fourth-grade students and 87 percent of eighth-grade students answered correctly, which suggests that most students in those grades could visualize what a folded piece of paper with a triangular piece cut on the fold would look like unfolded. However, students in grades 4 and 8 were less likely to select the correct answer for item 2, which asked which face of the cube would be on top if the flattened version were reassembled into a three-dimensional cube. These items suggest that students may have had ample experiences with symmetry, which is implicit in item 1, but not many experiences with visualizing how a flat object will appear if it is folded into a solid.

Table 7.3 Items Involving Visualization

Item	Percent Responding[a]		
	Grade 4	Grade 8	Grade 12

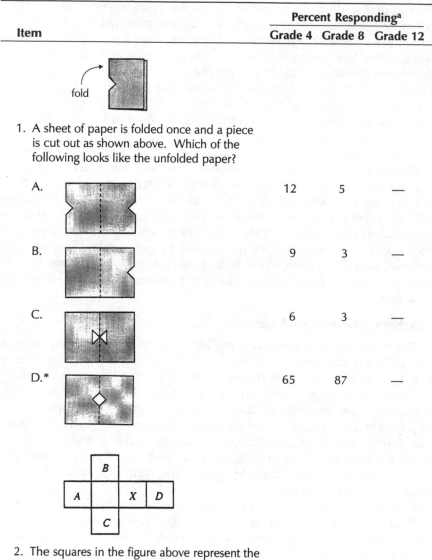

1. A sheet of paper is folded once and a piece is cut out as shown above. Which of the following looks like the unfolded paper?

A.	12	5	—
B.	9	3	—
C.	6	3	—
D.*	65	87	—

2. The squares in the figure above represent the faces of a cube which has been cut along some edges and flattened. When the original cube was resting on face X, which face was on top?

A. A*	22	55	—
B. B	28	14	—
C. C	16	8	—
D. D	30	22	—

Item	Percent Responding[a]		
	Grade 4	Grade 8	Grade 12

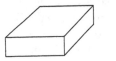

3. The piece of fudge above is in the shape of a rectangular solid. If a knife makes one straight cut through the fudge, which of the following can be the piece cut off?
Fill in one oval to indicate YES or NO for each shape.

○ YES   ○ NO

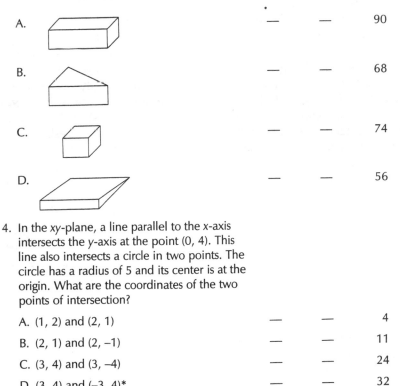

A.   —   —   90

B.   —   —   68

C.   —   —   74

D.   —   —   56

4. In the xy-plane, a line parallel to the x-axis intersects the y-axis at the point (0, 4). This line also intersects a circle in two points. The circle has a radius of 5 and its center is at the origin. What are the coordinates of the two points of intersection?

	Grade 4	Grade 8	Grade 12
A. (1, 2) and (2, 1)	—	—	4
B. (2, 1) and (2, –1)	—	—	11
C. (3, 4) and (3, –4)	—	—	24
D. (3, 4) and (–3, 4)*	—	—	32
E. (5, 0) and (–5, 0)	—	—	15

*Indicates correct response.
[a]In item 2, a multiple-response item, percents represent correct responses to each statement.
Note: Percents may not add to 100 because of rounding or omissions.

Examining the responses to item 2 revealed that a greater percentage of fourth-grade students chose B (the letter "on top" of the unfolded cube) than chose A (the correct answer) and even more chose D (the letter on top if the rightmost faces were folded). These results suggest that fourth-grade students were not visualizing three dimensionally at all, or perhaps visualizing only one fold. More eighth-grade students visualized better, as shown by the small percent who chose B, but another 22 percent were still visualizing only part of the problem. Students need more experience with different kinds of nets. Through folding nets students will not only improve their visualization skills but will also gain a better understanding of the relationship between three-dimensional and two-dimensional shapes.

For item 3 in table 7.3, twelfth-grade students were asked to visualize a three-dimensional rectangular solid cast in a real-world context as a piece of fudge and then to determine whether or not a particular shape could result from a single cut through the piece of fudge. Slightly more than 40 percent of twelfth-grade students correctly answered all four parts of the item, which suggests that some shapes, most likely the shapes resulting from diagonal cuts, were difficult for those students to identify. Ninety percent of the twelfth-grade students answered that the shape in A, which is also a rectangular solid, could be a piece cut from the larger piece of fudge. Recognizing that choices B and D represented shapes that would result from diagonal cuts was possibly more difficult, as shown by the lower percent-correct rates of 68 percent and 56 percent, respectively. Incorrect responses to choice C could have been a result of students' visualizing three dimensionally about a rectangular solid instead of reasoning solely from the figure shown and thereby concluding that the cube in C was possible if the rectangle on the front of the solid were a square. This would have led them to choose Yes for C, an incorrect response, despite their correct reasoning on the item. Clearly, this makes it hard to interpret students' performance on this item—it is not clear whether students chose Yes because they thought it could be produced even from the figure as it was given or whether they reasoned from the given information about the "rectangular solid" and concluded that a square front face was possible given that description.

Item 4 in table 7.3 is an example of an item that could have been solved by visualization or algebraically. Only 32 percent of the twelfth-grade students responded correctly. Although NAEP results provide no information on how students approached solving this problem, it is useful to think about possible strategies they could have used. One may assume that many of the students did not understand some of the terminology in the item and were unable to decode the problem in order to visualize it or solve it algebraically. However, if students had drawn a picture of the conditions stated in the problem, they would have seen that most of the choices could have been eliminated. Drawing the picture would have enabled the students to see that the coordinates (1, 2), (2, 1), (2, –1), (3, –4), (5, 0), and (–5, 0) were obviously below the line parallel to the x-axis that intersects the y-axis at point (0, 4). Thus, the only possible choice is D with coordinates (3, 4) and (–3, 4).

Furthermore, this problem could have been solved algebraically. Students could have found the equation of the circle ($x^2 + y^2 = 5^2$) and the equation of the line ($y = 4$) and used substitution to find the x-coordinates of the two points of intersection, but this is a much more difficult process and more prone to error. Even though this item is different from the other visualization item mentioned, it is significant because it shows the importance of encouraging students to interpret a problem visually rather than rely completely on algebraic methods.

While researchers have found positive correlation between spatial ability and mathematics achievement at all grade levels (e.g., Clements and Battista 1992), students' performance on NAEP items related to this area varies greatly depending on the relative complexity of the figures involved. Thus, students at all grade levels may benefit from more experience with activities that increase their spatial abilities. Students in primary through secondary grades can develop spatial sense through using tangrams, pentominoes, and tessellations; drawing and discussing geometric patterns; and constructing and comparing three-dimensional figures (Wheatley 1990). Students at higher grade levels can benefit from these experiences and others that require them to explore three-dimensional tasks.

IDENTIFICATION OF GEOMETRIC FIGURES AND THEIR ASSOCIATED PROPERTIES

Identification, description, comparison, and classification of geometric figures and their properties are also important skills in the content area of geometry (NCTM 1989). Recognizing and classifying geometric figures and understanding properties of geometric figures, such as symmetry and parallelism, are necessary both for students' use of geometric models to solve problems and for their development of an appreciation of the geometric nature of their physical world. Eleven NAEP items assessed students' ability to identify geometric figures or properties of geometric figures. Two items were administered only to fourth-grade students, two items only to eighth-grade students, and two items only to twelfth-grade students, with the remaining items administered at contiguous grade levels. Students in grades 4 and 8 were generally successful in identifying simple two-dimensional figures such as triangles and familiar three-dimensional figures such as cylinders, with percent-correct values ranging from 79 percent in grade 4 to 94 percent in grade 8. Results from items administered to eighth-

and twelfth-grade students show that as the geometric figures and the situations in the items became more complex, performance levels decreased.

Students in grade 4 were generally successful in identifying simple figures and their associated properties—for example, identifying a figure based on the number of angles or recognizing which figures were cylinders, with percent-correct values greater than 80 percent. However, performance levels dropped to below 45 percent for items involving more-complex concepts such as parallel lines and right angles. Although eighth-grade students had little trouble with right-angle concepts, they had difficulty identifying more-complex figures such as ellipses and special triangles and properties such as symmetry, with performance on items involving these complex concepts ranging from 48 percent to 32 percent correct. NAEP results for twelfth-grade students suggest that as the figures became more complex (for example, quadrilaterals as opposed to simple triangles), performance levels decreased.

As mentioned above, students in grade 4 had difficulty with the concept of parallel lines, as shown on their performance on a multiple-choice item that asked them to identify the letter that has two parallel lines, the choices being A, T, K, and N. Only 27 percent of the students selected N, with 42 percent selecting T. Although these results may be attributed to guessing, it could also be that fourth-grade students were confusing the terms *perpendicular* and *parallel*.

The items in table 7.4 illustrate the difficulties eighth- and twelfth-grade students had identifying geometric figures in complex problems, especially in situations in which the figure to be identified was embedded within another figure. For item 1, a regular constructed-response item, when eighth-grade students were asked to name the types of two-dimensional figures formed from perpendicular and oblique cross-sections of a cylinder, slightly less than half correctly identified all three figures and just over two-thirds identified the shape of two or more cuts. Compared to performance on item 1, performance on item 2, a multiple-choice item, was much lower. Only about one-third of the eighth-grade students correctly identified the intersection of the two triangles in a rectangle as a diagonal of that rectangle. In fact, a greater percent of students (36 percent) selected the distracter, "a right angle," than selected the correct answer, "a diagonal of the rectangle." However, this may have been due to a possible misinterpretation involving the term *intersection* as "what the figures have in common"; in particular, that $\triangle ABC$ and $\triangle ADC$ both contain right angles. Another possibility is that students misinterpreted the symbolism in the item; that is, they interpreted the triangle symbol ($\triangle$) as the symbol for an angle ($\angle$) and concluded that $\angle ABC$ and $\angle ADC$ were both right angles.

Item 3 in table 7.4 asked students to generalize about five figures, each of which was inscribed in a different quadrilateral. Only about half of the

Table 7.4 Identification of Geometric Figures: Sample Items

Item	Percent Responding	
	Grade 8	**Grade 12**

1. Each of the cylinders shown below was cut in a different way. The shaded part shows the shape of the cut. Under each figure, write the name of the shape of the cut.

Answer: _____   Answer: _____   Answer: _____

All three answers correct	48	—
One answer correct	8	—
Two answers correct	19	—
All three answers incorrect	17	—
Omitted	7	—

2. In the rectangle above, the intersection of △ABC and △ADC is which of the following?

A. A right angle	36	—
B. A vertex of the rectangle	15	—
C. A diagonal of the rectangle*	31	—
D. A pair of parallel sides	8	—
E. A pair of perpendicular sides	9	—

Continued

	Percent Responding	
Item	Grade 8	Grade 12

3. In the five quadrilaterals shown above, the midpoints of the sides have been joined by broken line segments. Which best describes the five dotted figures formed?

A. All are parallelograms.*	—	50
B. All are rectangles.	—	5
C. All are squares.	—	9
D. All are rhombuses.	—	10
E. No generalizations can be made.	—	24

*Indicates correct response.

Note: Percents may not add to 100 because of rounding or omissions.

students in grade 12 correctly concluded that the figures formed by the midpoints would all be parallelograms. Thus, NAEP results for item 3 in the table suggest that only about half of the students in grade 12 were functioning at the analytic level of the van Hiele levels (van Hiele 1986). Nearly one-fourth of the students selected the distracter associated with the idea that no generalizations can be made about the nature of the embedded figures. Because some of the figures formed were rectangles, some were rhombuses, some were squares, and some were parallelograms, students may have seen the five figures as different, even though they also may have seen that the figures had parallel sides. If the students did not see the parallel sides, they were exhibiting the "holistic" level of the van Hiele levels of geometric thought. At this level students do not focus on the explicit properties of shapes but see each shape as an object in a class of its own (Burger and Culpepper 1993). In this case the students did not see the common properties of the five figures. If the students did see that the figures had parallel lines but could not classify them all as parallelograms, then they were exhibiting the analytic level of geometric thought. At this level students are able to see figures in terms of their components and discover properties of a class of shapes but are unable to see how different classes of shapes overlap.

Identifying the common elements in a set of examples is critically important in geometry (and in all of mathematics), and students need to develop their abilities to reason inductively from sets of examples—gathering data and making observations, generating conjectures from those observations, and then attempting to prove those conjectures using deductive reasoning. This is another case in which the use of dynamic geometry tools such as the Geometer's Sketchpad and Cabri Geometry would enhance students' learning (Baulac, Bellemain, and Laborde 1992; Jackiw 1992).

Based on NAEP results, then, students at all three grade levels seemed to be quite successful in identifying and naming familiar two-dimensional and three-dimensional figures and their properties, operating at the holistic level of the van Hiele levels. However, when asked to identify figures in less familiar contexts or to recognize more-complex properties, students performed at lower levels, which suggests that students should have experience with a variety of geometric figures in more-complex contexts.

ANALYSIS OF PROPERTIES OF GEOMETRIC FIGURES

The nature of geometry—shapes and their properties and interrelationships—has made geometry a vehicle for developing mathematical reasoning skills (Geddes and Fortunato 1993). In order to develop a conceptual understanding of geometry, students need to be placed in situations that allow them to apply deductive, inductive, and spatial reasoning. Students must also be given experiences that enable them to make and evaluate conjectures and arguments to validate their own thinking (Geddes and Fortunato 1993). Ten NAEP items were designed to assess students' abilities to reason deductively, inductively, or spatially relative to geometric figures and their properties. Important concepts and processes addressed in the items included comparing geometric figures with respect to certain properties, using properties of particular figures to draw conclusions about them, identifying a figure as a counterexample to a given situation, and deducing generalizations that hold for all instances of a certain type of figure.

Students' performance on items in this area varied considerably; the items are described in table 7.5. More than 50 percent of students in grade 4 and 77 percent of students in grade 8 could identify a figure that provided an appropriate counterexample (item 1), more than 70 percent of students in grade 12 could choose a property that does not hold for rectangles (item 7), and about half of students in grade 12 could determine all four correct conclusions about segments produced in an angle bisector construction (item 8). This relatively high performance may have been influenced by the form in which the items were presented, that is, with figures that supported the reasoning process or that were presented as choices in a multiple-choice item. In addition, students may have reasoned from the figures instead of

Table 7.5 Performance on Items Requiring Reasoning about Geometric Properties

| | Percent Correct | | |
Item Description	Grade 4	Grade 8	Grade 12
1. Choose a figure that provides a counterexample for a false statement about quadrilaterals.	53	77	—
2. Compare geometric shapes and explain how they differ.	54	74	83
3. Compare geometric shapes and give examples of how they are similar and different.	10[a]	—	—
4. Choose a property that is not true for parallelograms.	—	41	64
5. Given a nondiameter chord and a diameter of a circle, select a statement that correctly relates their lengths.	—	33	—
6. Given several properties that define a particular geometric figure, select an additional property that must be true.	—	19	—
7. Choose a property that does not hold for rectangles.	—	—	71
8. Given a figure illustrating the construction of an angle bisector, determine whether each of four statements about the construction is true.	—	—	51[b]
9. From a set of triangles, choose one that does not have a particular property.	—	—	21
10. Determine whether a statement about regular polygons is true and explain.	—	—	18

[a]Percent of students scoring at either the satisfactory or extended level.
[b]Percent of students answering all four statements correctly.
Note: Item 3 was an extended constructed-response item, item 8 was a multiple-response item, items 2 and 10 were regular constructed-response items, and the rest were multiple-choice items.

basing their conclusions on their knowledge of geometric properties. This phenomenon of reasoning from figures has been alluded to earlier in the visualization section of this chapter and will be discussed in more detail in this section and others. Also, in these three problems the geometric figures were familiar ones, such as rectangles. For the other items involving less familiar figures, eighth- and twelfth-grade students' performance was not as high.

Items 1, 2, and 3 described in table 7.5 asked students to examine figures and determine similarities and differences. Students at all three grade levels did quite well on item 2, with 54 percent of the fourth-grade students, 74 percent of the eighth-grade students, and 83 percent of the twelfth-grade students answering correctly. These percents are similar to the percents for item 1, discussed above, which also contained simple geometric shapes. Students in grade 4, however, did not perform as well on the other comparison item, an extended constructed-response item worded in such a way as to encourage students to identify more than one way in which two quadrilaterals were the same and more than one way in which they differed. Only 10 percent of those students were able to describe at least one way the figures were the same and at least one way they differed. These results suggest that fourth-grade students can identify properties of simple geometric figures but have more difficulty working with more-complex figures and providing more-extensive written explanations. Students' performances on these problems exemplify that most students are at the holistic level of the van Hiele levels of geometric thought. Students are not reasoning at the analytic level, which would enable them to identify similarities and differences between geometric figures.

Three items (4, 6, and 7) in table 7.5 asked students to select which statements about properties of quadrilaterals were true or false. For item 4 in table 7.5, about two-fifths of students in grade 8 and nearly two-thirds of students in grade 12 correctly reasoned about properties of a parallelogram, a figure that is somewhat more complex than a rectangle or square. Because a figure was included in this item, one may assume that students may have reasoned from the figure. Thus, this item is congruent with the assumption that most students are operating on the holistic level of the van Hiele levels. Item 6 was more difficult for students; to answer it correctly students had to be operating at the informal deduction level of the van Hiele levels. In this particular item, students were given a set of statements about a geometric shape without a figure and asked to determine the correct statement on the basis of the given statements. Only 19 percent of the eighth-grade students answered this item correctly. The performance level on item 7 was relatively high, with about three-fourths of the twelfth-grade students responding correctly. However, one would have expected students to have done well on this particular item because it focused on properties of a rectangle, one of the most familiar geometric shapes.

The two items in table 7.6 illustrate the potential influence that figures have on students' reasoning processes. Often a figure can foster correct reasoning, but reasoning from the figure without regard to other information in the problem can lead to an incorrect conclusion. Item 1 in the table

Table 7.6 Reasoning about Geometric Figures: Sample Items

Item	Percent Responding[a]	
	Grade 8	**Grade 12**

1. Point *O* is the center of the circle above. Line segment *AC* is a diameter of the circle. Line segment *BC* does not pass through the center of the circle. Which of the following is true?

A. *AC* is longer than *BC*.*	33	—
B. *BC* is longer than *AC*.	9	—
C. *AC* and *BC* are the same length.	38	—
D. *BC* is twice as long as *OA*.	11	—
E. The lengths of *AC* and *BC* change, depending on how this piece of paper is turned.	6	—

2. The figure above shows the construction of the angle bisector of ∠ AOB using a compass. Which of the following statements must always be true in the construction of the angle bisector?

A. OA = OB	—	88
B. AP = BP	—	76
C. AB = BP	—	77
D. OB = BP	—	74

*Indicates correct response.

[a]In item 2, a multiple-response item, percents represent correct responses to each statement.

Note: In item 1, percents may not add to 100 because of rounding or omissions.

is an example of the latter situation, with item 2 an example of the former. In the figure accompanying item 1, the nondiameter chord BC is drawn nearly equal in length to the diameter AC, which possibly prompted some students to reason incorrectly from the diagram rather than from the information given in the narrative part of the problem. Performance on that item supports the notion that some eighth-grade students probably reasoned from the figure, since a higher percent of students (38 percent) chose the distracter associated with the line segments being equal than selected the correct answer (33 percent).

For item 2, a multiple-response item shown in the table, the figure can be helpful in deciding whether the statements are true or false. For example, segments OA and OB appear to be equal, and if the appropriate points are connected, segment AB appears to be shorter than segment BP. The high level of performance on the individual parts of item 2 shows that most twelfth-grade students were able to select the correct responses. However, it could be the case that students did not understand the properties related to constructing an angle bisector, despite correctly answering item 2. Instead, they may have reasoned directly from the figure. Giving items 1 and 2 to students and asking them to explain how they reasoned about their answers could provide teachers with insight into the extent to which students reasoned from the figure or reasoned from geometric properties.

NAEP results on items related to reasoning about geometric figures and their properties indicate that most students at all three grade levels are operating at the holistic level of the van Hiele levels of geometric thought. There is little evidence to support that many of them are operating at the higher levels. Thus, students need more experiences that allow them to use deductive, inductive, and spatial reasoning. The next section discusses performance on NAEP items that assessed how well students applied important geometric properties and concepts.

APPLICATION OF PROPERTIES OF GEOMETRIC FIGURES

NAEP items that required students to apply properties of geometric figures fell into two categories: those presented synthetically and those presented analytically. These two contexts represent different but complementary ways of representing geometric information. In synthetic contexts, students reason about figures in the absence of information about their locations, whereas in analytic contexts coordinates are given that place figures at particular locations in two or three dimensions. To solve analytic geometry problems, students use approaches such as slopes, distances between points, and locations of midpoints that draw heavily on the interplay between geometry and algebra. Because analytic and synthetic approaches are substantially different, it is important to examine students' performance in both contexts.

Synthetic Contexts

In NAEP, fourteen items involving application of properties of geometric figures were presented synthetically. Because only one item was administered to fourth-grade students, little can be concluded about those students' understandings of applying properties of geometric figures. Thus, this section will focus on the performance of students in grades 8 and 12 on the items summarized in table 7.7.

Table 7.7 Performance on Items Requiring Application of Properties: Synthetic Geometry

| | Percent Correct | |
Item Description	Grade 8	Grade 12
Triangles: Sum of Angles		
1. Determine the measure of one angle in a triangle, given the measures of the other two.	41	77
2. Determine the measure of a remote interior angle of a triangle.	32	—
3. Determine the measure of one angle in a triangle, given the measures of the other two.	—	70
Triangles: Pythagorean Theorem		
4. Given the length of two legs in a triangle, determine the length of the hypotenuse.	30	—
5. Given the length of two legs in a triangle, determine the length of the hypotenuse.	—	52
6. Sketch a right triangle based on given information about the lengths of the legs and the hypotenuse.	—	15
Circles and Polygons		
7. Determine the measure of a central angle in a circle.	31	—
8. Choose the correct value for the radius of an inscribed circle.	—	70
9. Describe a geometric process involving a circle.	—	12[a]

[a]Percent of students scoring at either the satisfactory or extended levels.

Note: Item 6 was a regular constructed-response item, item 9 was an extended constructed-response question, and the rest were multiple-choice items.

As shown in table 7.7, eighth-grade students' performance was relatively low, with percent-correct values ranging from 30 percent to 41 percent. The performance levels of the twelfth-grade students were a little better, ranging from 52 percent correct on an item involving the Pythagorean theorem to 77 percent correct on an item that asked students to determine the measure of an angle in a triangle, given the measure of the other two angles. However, students in grade 12 had trouble with complex problems involving triangles and circles.

Most, if not all, middle-grades mathematics curricula address the sum of the measures of the angles in a triangle. Despite that emphasis, the results for item 1 in table 7.7 show that only two-fifths of eighth-grade students could correctly find angle measures in a triangle based on knowledge of the sum of the angles. Results on item 2 for those students were even lower, with only about one-third selecting the correct answer. A potential explanation for the somewhat lower results on item 2 than on item 1 may be that students who had not explicitly encountered the exterior angle theorem for triangles approached the second item as a multistep problem involving supplementary angles and the triangle angle sum theorem. This may have been more complicated than the use of the triangle angle sum theorem in item 1, which would have required students only to subtract the sum of the measures of the two given interior angles from 180° to find the unknown interior angle measure.

On items involving angle sums in triangles, performance of twelfth-grade students was considerably better than that of eighth-grade students. About three-fourths of twelfth-grade students responded correctly to both item 1 and item 3, as described in table 7.7. For both grades, the most popular distracters for item 1 were one of the given angle measures or the sum of the two given angle measures, which suggests that students may have guessed one of the given measures or applied the "when in doubt, add" strategy to the two angle measures given. Nearly one-fourth of eighth-grade students chose one of the given angle measures, and another one-fourth chose the sum of the two given angle measures. By grade 12, however, fewer than 10 percent chose each of those distracters, which shows some growth in knowledge of properties of triangle-angle measures between grade levels.

Although the Pythagorean theorem is probably the most universally addressed theorem in geometry, there is evidence from NAEP that many students cannot apply it and probably do not understand it well. For example, for item 4 as described in table 7.7, fewer than one-third of eighth-grade students could find the length of the hypotenuse given lengths of the legs, despite all lengths being relatively small integers. In fact, of the four distracters, one distracter was selected by 34 percent of the students and another distracter by 26 percent. The first distracter represented the incorrect assumption that the hypotenuse and one leg were of approximately equal length. In fact, the figure that accompanied that item was drawn such that the legs' lengths were approximately equal. As was described in a previous section of this chapter, there is evidence in NAEP that some students reason incorrectly from figures, and performance on item 4 provides further evidence of this. With respect to the other distracter, representing the sum of the lengths of the legs, students who chose that distracter may have incomplete understanding of the Pythagorean theorem (for example, the sum of the leg lengths equals the length of the hypotenuse) or perhaps they defaulted again to a "when in doubt, add" strategy.

For students in grade 12, on a similar Pythagorean theorem item identified as item 5 in table 7.7, just over half chose the correct answer. One of the distracters for this item represented the sum of the leg lengths, but fewer than 20 percent of students in grade 12 selected that answer. This suggests that by grade 12, students are less likely than students in grade 8 to use the fallback strategy involving addition of the given numbers. The extremely poor performance of twelfth-grade students on item 6, a regular constructed-response item, is typical of performance levels for other such items. Because of the low level of performance on item 6 and its secure status, little additional information is available from the 1992 NAEP about twelfth-grade students' ability to work with the Pythagorean theorem in complex situations.

On items involving applications of geometric properties of circles and polygons, students appeared to be more successful when the item involved familiar content or procedures, such as finding the radius of a circle (item 8 in table 7.7), than when less-familiar content or procedures were involved. For example, on item 7 fewer than one-third of eighth-grade students were successful on an item involving the measure of a central angle in a circle, and on item 9, an extended constructed-response item, only 12 percent of twelfth-grade students scored at or above the satisfactory level. That extended item required students to apply properties of circles, and its difficulty was perhaps related to students' being required to describe a procedure and to explain why that procedure was correct. Additional performance data for that item reveal how difficult it was even for students in grade 12 who had taken more-advanced mathematics courses. Even among students who took mathematics beyond algebra 2, including those who took calculus, only about one-fifth scored at the satisfactory or extended levels.

The three released items in table 7.8 are examples of synthetic geometry items that required students to apply properties of regular polygons, angle bisectors, and similar triangles to find the measures of angles or sides in geometric figures. On item 1, 54 percent of eighth-grade students found the correct measure of an interior angle of a regular 24-gon when given the sum of the measures of the 24 interior angles. This level of performance,

Table 7.8 Items Set in Synthetic Contexts

Item	Percent Responding	
	Grade 8	Grade 12
1. The sum of the measures of the 24 angles in a a 24-sided regular polygon is 3,960°. What is the measure in degrees of one of the angles? Answer: _____		
Correct answer of 165° (or 165)	54	—
Any incorrect response	32	—
Omitted	15	—

Item	Grade 8	Grade 12
2. The sum of the measures of angles 1 and 2 in the figure above is 90°. What is the measure of the angle formed by the bisectors of these two angles?		
A. 60°	21	14
B. 45°*	23	44
C. 30°	27	21
D. 20°	17	12
E. 15°	8	6

Item	Grade 8	Grade 12
3. In the figure above, the two triangles are similar. What is the value of x? Answer: _____		
Correct answer of 19.6 or $9\frac{3}{5}$ or $\frac{48}{5}$	—	24
Incorrect answer of 6.66... or 6.7	—	2
Incorrect answer of 8	—	7
Any other incorrect answer	—	60
Omitted	—	8

*Indicates correct response.

Note: Percents may not add to 100 because of rounding or omissions.

even on a constructed-response item, may be deceiving, however, since students may not have needed to understand regular polygons to generate a correct answer. Without knowing the properties of regular polygons, a student could reason that since the sum of 24 items is 3,960, one would need to divide 3,960 by 24 to find the size of one of those items. This question may have tested little more than students' abilities to select a correct operation and perform a computation correctly.

Item 2 in table 7.8 required eighth- and twelfth-grade students to find the measure of the angle formed by the bisectors of two adjacent, complementary angles. Only 23 percent of eighth-grade students and 44 percent of twelfth-grade students answered correctly, and 30° was the most popular distracter. Students who chose 30° may have been giving their estimate of the measure of angle 2 from the figure, and those who selected 60° may have similarly been estimating the measure of angle 1. Thus, students who chose either 30° or 60° may have been reasoning incorrectly from the figure rather than reasoning from information about angle bisectors.

On item 3, which assessed the concept of similar triangles, fewer than one-fourth of twelfth-grade students found the correct side length. Although 69 percent of the students responding to this item answered it incorrectly, NAEP provides information only for two specific incorrect responses. Seven percent of the students chose 8 for the value of x, a value that most likely resulted from incorrectly reasoning from the figure that the segment with length x was approximately the same length as the segment with the given length of 8. Two percent gave an answer of 6.66... or 6.7, the value expected from incorrectly setting up the proportion as $6/5 = 8/x$, which suggests that a small percentage of students understood that in similar triangles corresponding sides are proportional but were simply unable to correctly identify corresponding sides.

Analytic Contexts

Three items, all administered at grade 12, requiring use of properties were presented in analytic contexts and involved slope, distance between points in the coordinate plane, and finding the coordinates of a vertex of a polygon. Performance levels ranged from 59 percent correct on the item involving finding coordinates for a vertex of a polygon to 33 percent on the item involving the use of the distance formula, which suggests that students in grade 12 have a somewhat uneven understanding of concepts and procedures associated with analytic geometry.

The two released items in table 7.9 illustrate the difficulties students had with analytic geometry items. On item 1 only 41 percent of the students answered correctly. Another 15 percent of students selected the distracter associated with the reciprocal of the correct response, which suggests that those students may have found the horizontal and vertical change but divided incorrectly. The most popular distracter, 1/3, most likely resulted from a signed-number mistake in the point-slope formula. On item 2, only one-third of twelfth-grade students correctly chose 10 as the distance, and of those who did, some may have obtained that response simply by adding the four given coordinates $(2 + 10 - 4 + 2 = 10)$. Several things may have contributed to item 2 being somewhat more difficult than item 1, in particular, the lack of a figure and students' need to correctly

Table 7.9 Items Set in Analytic Contexts

Item	Percent Responding Grade 12

1. What is the slope of the line shown in the graph above?

A. $\frac{1}{3}$	18
B. $\frac{2}{3}$ *	41
C. 1	15
D. $\frac{3}{2}$	14
E. 8	8

2. What is the distance between the points (2, 10) and (−4, 2) in the xy-plane?

A. 6	14
B. 8	23
C. 10*	32
D. 14	20
E. 18	3

*Indicates correct response.

Note: Percents may not add to 100 because of rounding or omissions.

recall the distance formula. In item 2, had students been given a figure in which the given points were plotted, they may have recognized that the segment connecting the two points was the hypotenuse of a right triangle that could be used to find the distance between the points.

NAEP results show that some students are able to apply geometric properties in routine situations. For example, students who knew that the sum of the measures of the interior angles of a triangle is 180 degrees could use a straightforward procedure to find the answers for the "triangle angle sum" items (for example, items 1 and 3 in table 7.7). Similarly, finding the length of a right triangle's hypotenuse, given the length of the two legs (items 4 and 5 in table 7.7) and determining the slope of a line given two points (item 1 in table 7.9), were essentially procedural tasks if students knew the correct formulas. However, many of the items requiring application of properties of geometric figures did not substantially address important aspects of mathematics suggested in the NCTM *Curriculum and Evaluation Standards*, namely, problem solving, reasoning, communication, and connections. Unfortunately, NAEP results concerning the extent to which students possess knowledge of a theorem and the ability to recognize that it applies in a particular item gives mathematics educators very limited information about the extent to which students can do mathematics in the spirit of the *Standards*.

OVERALL PERFORMANCE IN GEOMETRY

This chapter has discussed student performance on individual geometry items and clusters of related items. Gaining a more global view of student performance for geometry as a NAEP content area requires looking at performance across the entire set of NAEP geometry items based on the NAEP scale. Table 7.10 presents data on the average performance in geometry for students in grades 4, 8, and 12 for 1990 and 1992 and includes data for selected subgroups.

Overall performance in geometry significantly increased from 1990 to 1992 for all three grade levels. In both years, performance increased steadily from grade 4 to grade 8 and from grade 8 to grade 12.

For males and females, performance levels in geometry were nearly identical in 1990 and 1992. Females at all three grade levels performed at a significantly higher level in 1992 than in 1990; males at grades 4 and 12 performed better in 1992 than 1990. For the race/ethnicity subgroups, White students' performance in geometry increased from 1990 to 1992 at grades 4, 8, and 12. For Black and Hispanic students the increases in performance were significant only at grades 4 and 12. Based on the 1992 results, there were disparities in performance in geometry between race/ethnicity subgroups: White students scored significantly higher than both Black and

Table 7.10 Average Performance in Geometry: Overall Performance and Performance by Gender and Race/Ethnicity, Grades 4, 8, and 12 for 1990 and 1992

	Grade 4	Grade 8	Grade 12
Overall			
1992	221 (0.7)>	263 (0.9)>	300 (1.0)>
1990	213 (0.9)	260 (1.3)	295 (1.3)
Gender			
Male			
1992	222 (0.8)>	263 (1.2)	304 (1.2)>
1990	213 (1.2)	261 (1.6)	298 (1.5)
Female			
1992	220 (0.9)>	263 (1.0)>	298 (1.2)>
1990	213 (1.2)	259 (1.3)	293 (1.6)
Race/Ethnicity			
White			
1992	229 (0.8)>	272 (1.1)>	306 (1.1)>
1990	220 (1.1)	267 (1.4)	302 (1.5)
Black			
1992	196 (1.4)>	234 (1.7)	276 (1.9)>
1990	191 (1.8)	236 (3.1)	267 (2.1)
Hispanic			
1992	206 (1.3)>	246 (1.2)	286 (2.5)>
1990	199 (1.9)	243 (2.5)	276 (2.9)

> The value for 1992 was significantly higher than the value for 1990 at about the 95 percent confidence level. The standard errors of the estimated proficiencies appear in parentheses.

Hispanic students in all three grades, and Hispanic students scored significantly higher than Black students in grades 4 and 8 only.

CONCLUSION

As shown throughout the chapter, students exhibit a surface understanding of most geometric concepts and skills. Furthermore, students in grades 4 through 12 are operating primarily at the holistic level of the van Hiele levels. For example, fourth-grade students were able to identify properties of simple geometric figures but had great difficulty working with more-complex figures and providing written explanations. The Pythagorean theorem is probably the most universally addressed theorem in geometry, but there is evidence from NAEP that many students cannot apply it and probably do not understand it well. Based on the 1992 NAEP mathematics assessment, students' school mathematics experiences with geometry must change if students are to attain the level of geometric understanding called for by the National Council of Teachers of Mathematics (1989) in the *Curriculum and Evaluation Standards for School Mathematics*. Students need more experiences with concrete models to enhance their visualization skills, more opportunities to explore the properties of geometric shapes and draw their own conclusions, and more opportunities to see how geometric concepts relate to real-life situations and other mathematical contexts.

REFERENCES

Baulac, Yves, Franck Bellemain, and Jean-Marie Laborde. Cabri: The Interactive Geometry Notebook. Pacific Grove, Calif.: Brooks/Cole, 1992.

Burger, William F., and Barbara Culpepper. "Restructuring Geometry." In *Research Ideas for the Classroom: High School Mathematics*, edited by Patricia S. Wilson, pp. 140–54. New York: Macmillan, 1993.

Clements, Douglas H., and Michael T. Battista. "Geometry and Spatial Reasoning." In *Handbook of Research on Mathematics Teaching and Learning*, edited by Douglas A. Grouws, pp. 420–64. New York: Macmillan, 1992.

Geddes, Dorothy, and Irene Fortunato. "Geometry: Research and Classroom Activities." In *Research Ideas for the Classroom: Middle Grades Mathematics*, edited by Douglas T. Owens, pp. 199–222. New York: Macmillan, 1993.

Jackiw, Nicholas. The Geometer's Sketchpad. Berkeley, Calif.: Key Curriculum Press, 1992.

National Assessment of Educational Progress. *Mathematics Objectives: 1990 Assessment*. Princeton, N.J.: Educational Testing Service, National Assessment of Educational Progress, 1988.

National Council of Teachers of Mathematics. *Curriculum and Evaluation Standards for School Mathematics*. Reston, Va.: National Council of Teachers of Mathematics, 1989.

Olive, John. "Learning Geometry Intuitively with the Aid of a New Computer Tool: The Geometer's Sketchpad." *The Mathematics Educator* 2 (Summer 1991): 26–29.

van Hiele, Pierre M. *Structure and Insight: A Theory of Mathematics Education.* Orlando, Fla.: Academic Press, 1986.

Wheatley, Grayson H. "Spatial Sense and Mathematics Learning." *Arithmetic Teacher* 37 (February 1990): 10–12.

8

What Do Students Know about Data Analysis, Statistics, and Probability?

Judith S. Zawojewski & David S. Heckman

D ATA ANALYSIS, statistics, and probability are important mathematical top-ics for a society based on technology and communication. This chap-ter provides information on what students know about collecting and or-ganizing data and making predictions on the basis of data and about concepts in statistics and probability. The first section briefly discusses data analysis, statistics, and probability as a content area in NAEP and describes the corresponding set of items. The next section contains information about student performance on the data representation, statistics, and probability items. The final section contains information on overall performance for students in this NAEP content area.

DATA ANALYSIS, STATISTICS, AND PROBABILITY IN THE NAEP FRAMEWORK AND ITEMS

In NAEP this content area "focuses on data representation and analysis across all disciplines, [with emphasis on] appropriate methods for gather-ing data, the visual exploration of data, and the development and evalua-tion of arguments based on data analysis" (National Assessment of Educational Progress 1988, p. 26). The percent distribution of NAEP items on data analysis, statistics, and probability was 10 percent at grade 4 and 15 percent at grades 8 and 12. Particular topics assessed at all three grade levels in NAEP included reading and interpreting tables and graphs, orga-nizing and displaying data in tables and graphs and making inferences, and determining probabilities for simple events. In grades 8 and 12 the list of topics was expanded to include those associated with measures of cen-tral tendency and dispersion, the recognition of randomness and bias in data collection, and the use and misuse of statistics in real-world applica-tions. Advanced concepts such as fitting a line or curve to a set of data and making predictions and the use of formulas were assessed at grade 12.

The 1992 NAEP mathematics assessment contained fifty-three items dealing with data analysis, statistics, and probability. A subset of items was administered at more than one grade level to facilitate comparisons among grade levels. There was about the same number of multiple-choice items as constructed-response items; three constructed-response items were extended questions that required students to explain their work and reasoning processes in detail. For some data analysis, statistics, and probability items students were permitted to use calculators.

An analysis of the data analysis, statistics, and probability items reveals that the number of items on data representation and interpretation outnumbered those dealing with topics in statistics and probability. Twenty-eight items involved reading or interpreting data from tables and graphs or creating tables and graphs for a given set of data. About one-third of the remaining twenty-five items assessed probability concepts ranging from the analysis of simple events at grade 4 to more complex, problem-solving situations at grade 12. The statistics items dealt primarily with the topics of average (mean) and median. Mode as a measure of central tendency was not assessed, and few items assessed students' understandings of common misconceptions about statistics and probability. The next section of this chapter includes a discussion of students' performance on individual NAEP data analysis, statistics, and probability items or clusters of related items in the following areas: data representation and interpretation, probability, and statistics.

HIGHLIGHTS

- About half the fourth-grade students and most eighth- and twelfth-grade students can read, use, and complete data entries in and from tables, pictographs, and bar graphs.
- About three-fourths of the twelfth-grade students can read and use data from line graphs.
- There is confusion about the meaning of the measures of central tendency, especially that of median, for eighth- and twelfth-grade students.
- More than half the fourth-grade students can work with probability concepts in which the probability involves one favorable outcome out of a total number of outcomes.
- Students in all three grade levels appear to have difficulty communicating their reasoning about data representation when responding to regular and extended constructed-response items.
- Performance levels in data analysis, statistics, and probability increased significantly from 1990 to 1992 for students in grades 8 and 12. (No comparison for fourth-grade students was possible because no performance scale was created in 1990 for students in that grade.)

READING AND INTERPRETING DATA

Reading, interpreting, and using data from tables and graphs are important skills for all, as people encounter various data representations on television and in newspapers, magazines, and reports for work or school. The NCTM *Curriculum and Evaluation Standards for School Mathematics* (1989) emphasizes the importance of reading and interpreting data from tables, graphs, and other representational formats throughout the K–12 curriculum.

The 1992 NAEP included twenty-eight items that involved reading, interpreting, and representing data in tables and graphs. First analyzed is student performance on items in which data are represented in tables, pictographs, and bar graphs. This analysis is followed by an examination of student performance on items involving line graphs and scatterplots.

Tables, Pictographs, and Bar Graphs

Tables, pictographs, and bar graphs are often introduced into the mathematics curriculum in the third or fourth grade. Reflecting that curricular emphasis, the NAEP mathematics assessment included nine items that asked fourth-grade students to read and interpret data from tables and graphs or required students at that grade to construct and complete graphs on the basis of given data. A subset of the grade-4 items was also administered to students in grade 8 and in grade 12 to facilitate comparisons across grade levels. Other items were also administered only to students in grade 8 or in grade 12 to assess advanced data-interpretation concepts such as the use and misuse of representations and data interpolation or extrapolation.

Interpreting and Using Data

Five items assessed fourth-grade students' ability to interpret and use data from tables and graphs. In three items the data were represented in tables; in the other two items the data appeared in the form of a pictograph in one item and a bar graph in the other. Results from those items indicate that students in grade 4 were generally successful in interpreting and using data presented in a variety of ways, with percent-correct values ranging from 48 percent to 67 percent.

The items in table 8.1 illustrate student performance on those items. Items 1, 2, and 3 were a set of related items, which asked students to obtain particular data from the table and then perform arithmetic operations with the data to answer the question posed in the item. For those three items students were provided with, and permitted to use, simple, four-function calculators. Among the three items, fourth-grade students had the highest

Table 8.1 Interpreting and Using Data from Tables and Graphs: Sample Items

	Percent Correct	
Item	Grade 4	Grade 8

Items 1, 2, and 3 below refer to the following table:

POINTS EARNED FROM SCHOOL EVENTS

Class	Mathathon	Readathon
Mr. Lopez	425	411
Ms. Chen	328	456
Mrs. Green	447	342

Item	Grade 4	Grade 8
1. Which class earned the most points from the two events? [Answer: Mr. Lopez's class]	67	—
2. What was the total number of points earned from the mathathon? [Answer: 1200]	52	—
3. Ms. Chen's class earned how many more points from the readathon than from the mathathon? [Answer: 128]	49	—

AGES OF CHILDREN IN
MR. RIVERA'S CLASS

Item	Grade 4	Grade 8
4. The graph above shows how many of the 32 children in Mr. Rivera's class are 8, 9, 10, and 11 years old. Which of the following is true? [Answer: Most are 9 or older.]	48	62

Note: Items 1 and 4 were multiple-choice items, and the rest were regular constructed-response items. For items 1, 2, and 3, students were provided with, and permitted to use, simple, four-function calculators.

percent correct on item 1, a multiple-choice item that required adding numbers in each column and comparing the sums; those students had the lowest percent correct on item 3, a regular constructed-response item that involved selecting data from the correct row and then subtracting.

Data linking the response to the self-report question on calculator use ("Did you use the calculator on this question?") to performance levels reveals that for each of the three items, students who reported using the calculator had significantly higher percent-correct values than students who reported not using the calculator. For example, for item 3 described above, the percent-correct values for students who reported using the calculator and those who reported not using the calculator were 62 percent and 29 percent, respectively. While it may be that students who chose to use calculators are those who have confidence in what procedures to perform, it may also be that the calculator helps to increase students' accuracy in computations with multidigit numbers. For item 4, a multiple-choice item that asked students to decide which statement was supported by information in the bar graph, performance was also low and suggests that fourth-grade students have difficulty reasoning about, and making comparisons using, data. By grade 8, however, 62 percent of students answered correctly, which shows that middle school students appear to have more facility in reasoning from data and in making or drawing inferences.

Constructing or Completing Tables and Graphs

Eight secure NAEP items, described in table 8.2, assessed students' ability to construct or complete tables and graphs from numerical data. Students in grade 4 were generally successful completing bar graphs and pictographs, especially when the given data could be used directly. For example, for item 1 and item 4, which required students to construct a bar graph and a pictograph, respectively, about 60 percent of the fourth-grade students drew correct graphs. Success rates were somewhat lower when students had to derive the information needed to complete the graph. For example, item 3 required that students derive information by first analyzing a multiplicative relationship among the categories being represented in the graph. Here, only 33 percent of the students in grade 4 drew a completely correct graph, with another 23 percent drawing one bar correctly. The multiplicative relationship among the categories was a likely source of difficulty on that item for fourth-grade students. In fact, results on a multiple-choice item, classified by NAEP as numbers and Operations, that involved a multiplicative relationship among groups of students were very similar to those for item 3 in table 8.2; that is, for the numbers and operations item, only about one-third of the fourth-grade students answered correctly. The results for the data analysis item, which was also administered to eighth-grade students, may suggest the emergence of proportional rea-

Table 8.2 Performance on Items Involving Constructing or Completing Tables and Graphs

Item Description	Percent Correct		
	Grade 4	Grade 8	Grade 12
Bar Graphs			
1. Use given data to complete a graph.	64	—	—
2. Use given data directly from a table to complete a graph.	51	85	91
3. Derive data from a given multiplicative relationship and complete a graph.	33	78	—
Pictographs			
4. Use given data directly to construct a pictograph using the appropriate number of icons.	59	90	—
5. Derive data from a table and construct a pictograph using the appropriate number of icons.	42	—	—
6. Use data directly from a table to choose the correct number of icons needed to represent a table entry.	—	78	—
Tables			
7. On the basis of given information, determine where to place a new entry in a table.	—	—	83
8. Use information from a table to determine the appropriate form for a new entry in a table.	—	—	70

Note: Item 6 was a multiple-choice item, and the rest were regular constructed-response items.

soning in middle school in that almost 80 percent of those students drew the correct graph, with another 10 percent drawing at least one bar correctly.

At the upper grade levels, success in constructing or completing graphs and tables increased, with more than three-fourths of the eighth-grade students responding correctly to these items. Notable was the disparity in performance by eighth-grade students on items 4 and 6 in the table: the two items were similar in that neither required the derivation of data to answer the question posed. The performance differences probably cannot be attributed to the formats of items in the same way as for other pairs of multiple-choice and regular constructed-response items. The usual pattern generally involves higher performance on multiple-choice items than on constructed-response items, but for items 4 and 6, the reverse is true. Perhaps some of the performance differences can be attributed to the scale used in each item. In particular, the scale in the constructed-response item involved one icon representing a one-digit number of objects, and the scale in the multiple-choice item involved one icon representing a three-digit number of objects. Using the latter scale may have been a more demanding calculation for students in grade 8 and may explain a lower performance level.

Problem Solving Using Data

Some NAEP data analysis items required students to use problem-solving skills to reason about data and then to present their responses in writing. The next sections focus on two of these items: a regular constructed-response item referred to hereafter as Trash Cans and an extended constructed-response item referred to as Graphs of Pockets.

The item called Trash Cans, shown in table 8.3, required eighth-grade students to interpret a graph holistically in order to explain why it was misleading. To answer correctly, students needed to focus on the fact that the graphic was intended to be a bar graph or pictograph, where the height of the 1980 trash can is twice that of the 1960 can, representing a doubling of the tonnage of trash between 1960 and 1980. In this graph, however, because both the height and width of the 1980 can were doubled, the visual impact of the quadrupled area distorts the intended relationship between 1960 and 1980. Technically, the actual trash can for 1980 depicted in the graph would hold eight times as much garbage as the 1960 can. According to the scoring guide, which was based on two levels—right or wrong—a technically correct response had to mention that both the width and the height of the 1980 can have been doubled, or that the 1980 can holds much more than twice the amount of the 1960 can, or that the ratio of the amount in the 1960 can to the amount in the 1980 can is less than one-half. An example of a correct response follows: "Both the width and the height of the 1980 can have been doubled. Only the height should have been doubled" (Dossey, Mullis, and Jones 1993, p. 37). Only 8 percent of the students in grade 8 responded correctly to this item, 86 percent responded incorrectly, and 6 percent left their papers blank.

This item and the student performance associated with it were described in an NAEP publication (Dossey, Mullis, and Jones 1993), but only one

Table 8.3   Trash Cans Question

Item	Percent Responding
	Grade 8

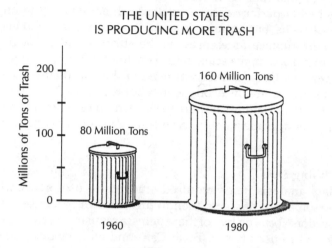

THE UNITED STATES
IS PRODUCING MORE TRASH

The pictograph shown above is misleading. Explain why.

Answer: _____

_____

_____

Correct response	8
Any incorrect response	86
Omitted	6

example of a correct response, quoted in the paragraph above, and no examples of incorrect responses were included. To get a better sense of how students were thinking about this problem, a set of approximately a hundred student responses, representing almost 10 percent of the total number of responses, was obtained and analyzed for this chapter. The primary intent of the analysis was to gain some insight into the errors and misconceptions evident in students' responses. The analysis identified some responses in which students showed a partial understanding of why the graph was misleading but did not explain their thinking completely. In other responses, some students criticized the technical aspects of the graph, others posited that necessary information was not included in the graph, and still others gave explanations based on their knowledge of real-world situations. Examples of these responses appear in figure 8.1.

The first example in the figure, labeled response 1, represents an incomplete response in which students referred to the relative sizes in general ways or only noted that one dimension was doubled. For some of these students, it may be that their lack of mathematical communication skills hindered their efforts to reveal the mathematics they knew, which underscores the need to provide students with opportunities to communicate mathematically. In particular, such students need to learn in their everyday classroom experiences what it means to communicate a *complete* response. Response 2 represents another kind of incomplete response found in the sample. Here, the student possibly was visualizing four of the small trash cans covering the picture of the large can. Although the visual representation of the trash cans is quadrupled, the increase in size of the actual trash cans would be eightfold. In fact, using the Trash Cans problem as a basis for a classroom investigation of the relationship between the fourfold increase in the area of the representation and the eightfold increase in the volume of the trash can might be an interesting activity.

Responses 3 and 4 in the figure suggest that some students were distracted by what they considered questionable graphing techniques and focused on a relatively less significant aspect of the graph. In response 3, the student appears to be preoccupied with how to read accurately the height of the larger can when the bottom of that can appears to be below the horizontal axis. The markings on the pictograph in response 4 illustrate the literal way in which the students tried to obtain exact measures for the height of each can. In particular, the students drew additional ticks on the vertical axis, which made it more detailed, and then drew a horizontal line from the top of each trash can to the vertical axis in an apparent attempt to measure precisely the actual height of the can. Although in each response the students' technical remarks have some validity, their overconcern with "technique," accompanied by the lack of knowledge about conventions in displaying data (e.g., using the center of the lid as the height), probably prevented a mathematically based explanation for why the graph was misleading. Distinguishing between minor technical inaccuracies and potential major misrepresentation of data needs to be addressed explicitly in the curriculum if students are to recognize and understand the impact of each.

Another type of incorrect response involved students' concern with missing information, which was not actually needed to respond adequately to the prompt. For example, in response 5 and response 6, some students thought that more years needed to be represented to make the graph more accurate. Finally, some responses reflected students' tendency to interpret the problem in light of real-world situations. Response 7 represents a class

Response 1

THE UNITED STATES
IS PRODUCING MORE TRASH

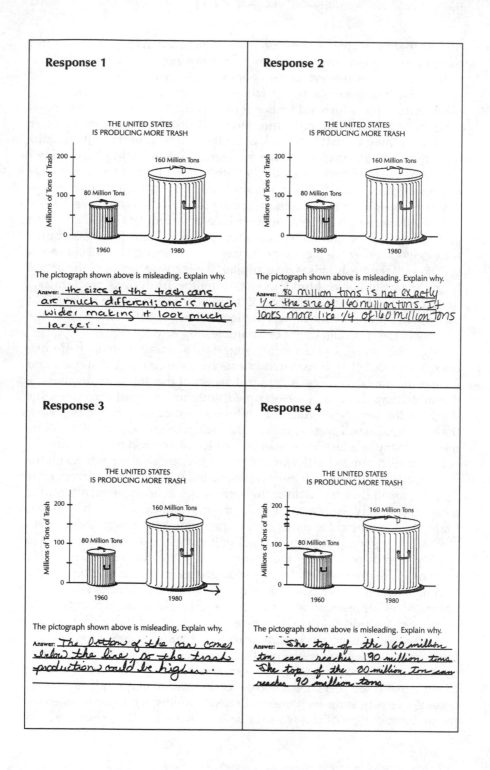

The pictograph shown above is misleading. Explain why.

Answer: the sizes of the trash cans are much different; one is much wider making it look much larger.

Response 2

THE UNITED STATES
IS PRODUCING MORE TRASH

The pictograph shown above is misleading. Explain why.

Answer: 80 million tons is not exactly 1/2 the size of 160 million tons. It looks more like 1/4 of 160 million tons

Response 3

THE UNITED STATES
IS PRODUCING MORE TRASH

The pictograph shown above is misleading. Explain why.

Answer: The bottom of the can comes below the line so the trash production could be higher.

Response 4

THE UNITED STATES
IS PRODUCING MORE TRASH

The pictograph shown above is misleading. Explain why.

Answer: The top of the 160 million ton can reaches 190 million tons. The top of the 80 million ton can reaches 90 million tons.

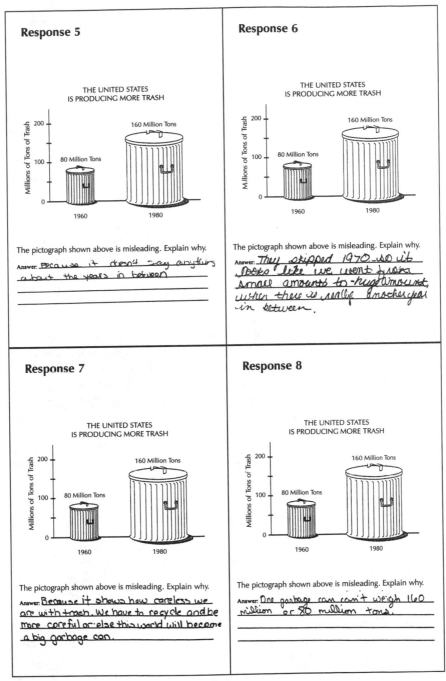

Response 5

THE UNITED STATES
IS PRODUCING MORE TRASH

Millions of Tons of Trash

200

160 Million Tons

100

80 Million Tons

0

1960   1980

The pictograph shown above is misleading. Explain why.

Answer: Because it doesn't say anything about the years in between

Response 6

THE UNITED STATES
IS PRODUCING MORE TRASH

Millions of Tons of Trash

200

160 Million Tons

100

80 Million Tons

0

1960   1980

The pictograph shown above is misleading. Explain why.

Answer: They skipped 1970 so it looks like we went from small amounts to high amount, when there is really Another year in between.

Response 7

THE UNITED STATES
IS PRODUCING MORE TRASH

Millions of Tons of Trash

200

160 Million Tons

100

80 Million Tons

0

1960   1980

The pictograph shown above is misleading. Explain why.

Answer: Because it shows how careless we are with trash. We have to recycle and be more careful or else this world will become a big garbage can.

Response 8

THE UNITED STATES
IS PRODUCING MORE TRASH

Millions of Tons of Trash

200

160 Million Tons

100

80 Million Tons

0

1960   1980

The pictograph shown above is misleading. Explain why.

Answer: One garbage can can't weigh 160 million or 80 million tons.

Fig. 8.1   Examples of responses to Trash Cans

of responses that dealt with the need to recycle, and response 8 is typical of the set of responses in which students interpreted the trash can in a literal sense; that is, they doubted the credibility of the situation depicted in the Trash Cans problem because no can could hold that much trash.

In light of the fact that graphics similar to that in the Trash Cans problem are seen daily in many places, including in newspapers and magazines and on television, students need to develop knowledge about various representations of data that goes far beyond knowing the techniques of constructing a table or graph. Thus, it would be useful if students were presented opportunities in school to evaluate the connections among actual data, the representation of the data, and the real-world situations being represented. Through such experiences, students can learn to analyze data critically and draw valid conclusions.

Another item on NAEP that assessed problem solving using data was the Graphs of Pockets question, shown in table 8.4. This question required that fourth-grade students read, interpret, and select one of three pictographs that best represents the situation described in words and based on two constraints: (1) there were 20 students in Mr. Pang's class; and (2) most of those students had pockets in their clothes. To answer the problem completely, students not only had to use both constraints to explain why the graph they chose was correct but also had to justify why they rejected the other two graphs. Unlike the Trash Cans problem, which was scored according to a right-or-wrong scoring scheme, the Graphs of Pockets problem was scored according to a focused holistic scheme based on five performance categories—incorrect, minimal, partial, satisfactory, and extended—and the designation "no response" for blank papers (also called "omitted"). Figure 8.2 contains descriptions of the criteria for each of the five performance categories and an illustrative example of a response from each category. The descriptions and examples are as they appeared in an NAEP publication (Dossey, Mullis, and Jones 1993, pp. 109–11).

Results for Graphs of Pockets, shown in table 8.4, reveal that fourth-grade students had difficulty with this question. Only 10 percent of the students produced a satisfactory or extended response, and almost half gave responses that were completely incorrect or irrelevant or that stated, "I don't know." The 38 percent of students whose responses were scored as minimal or partial were judged to have some understanding of the problem because their choice or their explanation "showed a grasp of the relationship between the problem situation expressed in words and the three graphs" (Dossey, Mullis, and Jones 1993, p. 112). However, a common difficulty in responses at those two score levels involved the ability to deal simultaneously with the dual constraints in the problem. For example, in the minimal response shown in figure 8.2 the student chose graph A and provided an explanation based on the number of students and did not

Table 8.4 Graphs of Pockets Question

	Percent Responding
Question	**Grade 4**

[General directions] Think carefully about the following question. Write a complete answer. You may use drawings, words, and numbers to explain your answer. Be sure to show all of your work.

There are 20 students in Mr. Pang's class. On Tuesday most of the students in the class said they had pockets in the clothes they were wearing.

A

Number of Pockets (y-axis, 0–10)

♀ = 1 Student

B

Number of Pockets (y-axis, 0–10)

♀ = 1 Student

C

Number of Pockets (y-axis, 0–10)

♀ = 1 Student

Which of the graphs most likely shows the number of pockets that each child had? _____

Explain why you chose that graph.

Explain why you did not choose the other graphs.

Extended response	3
Satisfactory response	7
Partial response	15
Minimal response	23
Incorrect	46
Omitted	6

Incorrect—The work is completely incorrect or irrelevant, or the response states, "I don't know."

C

because they are
all equal

because they are not
equal

Minimal—Student chooses Graph B with no explanation or the student chooses Graph A or Graph C with an explanation that shows some understanding.

A

I did because A had
20 students.

I did because they had
more than 20 students.

Partial—Student chooses Graph B but does not give an adequate explanation or student chooses Graph B but gives no explanation why, but explains why the answer is neither Graph A nor Graph C.

B

I chose B because
most of the people had
pockets.

I didn't choose the other
graphs because more people
didn't have pockets.

Satisfactory—Student chooses Graph B and gives a good explanation but does not mention the other graphs, or student gives a good explanation of why the answer cannot be Graph A or Graph C, but does not give a good explanation of why the answer is Graph B.

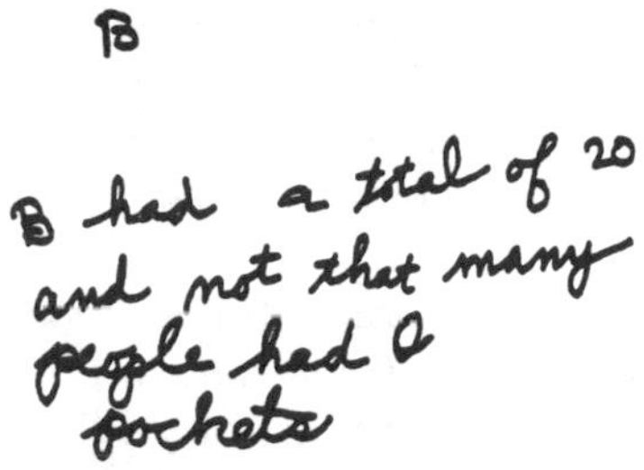

ß

B had a total of 20 and not that many people had a pockets

Extended—Student chooses Graph B, explains why the answer must be Graph B, and explains why neither Graph A nor Graph C can be the correct solution.

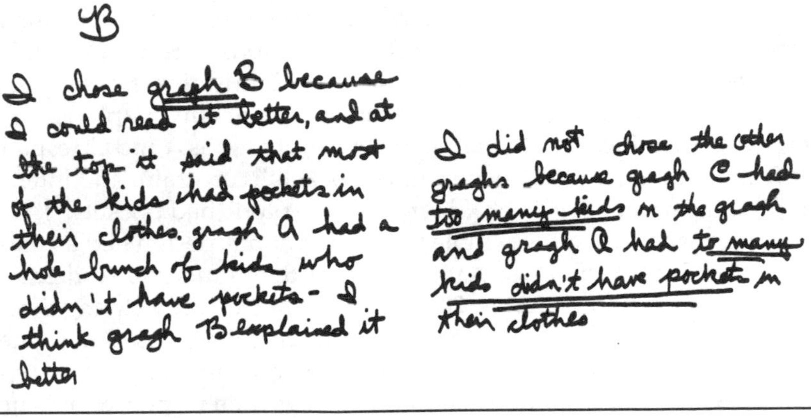

ɣß

I chose graph B because I could read it better, and at the top it said that most of the kids had pockets in their clothes. graph A had a whole bunch of kids who didn't have pockets- I think graph B explained it better

I did not chose the other graphs because graph C had too many kids in the graph and graph A had to many kids didn't have pockets in their clothes

Fig. 8.2 Graphs of Pockets: Performance categories and sample responses
Source: Dossey, Mullis, and Jones 1993

consider the "most of them have pockets" constraint; in the partial response shown in the figure, the student chose graph B but gave an inadequate explanation based on the constraint that "most people had pockets," never mentioning the other constraint of 20 students.

Although some information can be gleaned from the performance criteria and the five sample responses, such as the difficulty in dealing with the dual constraints just explained, an examination of a larger set of responses can reveal more information about students' errors and misconceptions about the Graphs of Pockets problem. To that end, a sample of responses representing about 45 percent of the total number of nonblank responses was obtained and analyzed for this chapter. Of particular interest were responses that designated a specific graph (graph A, graph B, or graph C) and included an explanation about why that graph most likely represented the distribution of pockets in Mr. Pang's class. This analysis confirmed the difficulty students had in dealing with the dual constraints in the question and revealed at least two other areas of difficulty: interpreting a graph correctly and writing complete and clear mathematically based justifications for their choices. Figure 8.3 contains additional examples of responses that exemplify one or both of these areas of difficulty.

Of the responses in the sample designating graph B as the correct graph, some focused on both constraints (20 children and most had pockets), with others either mentioning only the number of students or only the fact that most students had pockets. This pattern of responses suggests that even though fourth-grade students could identify the correct graph, they had difficulty attending to both constraints simultaneously. With respect to the justifications about why graphs A and C were inappropriate, not all responses contained clear explanations, and some responses contained incorrect interpretations of one or both of the graphs. Responses 1 and 2 in figure 8.3 are presented as illustrations of the kinds of responses just discussed. In response 1 there is a clear explanation of why graph B is correct and a vague justification for the incorrectness of graphs A and C; response 2 focuses only on the "most people have pockets" constraint and interprets graph C incorrectly ("[Graph C] said that everyone had pockets.").

Of the responses in the sample that had graph A designated as the correct graph, most responses focused only on the number of students, correctly stating that graph A has 20 students but incorrectly stating that *both* the other two graphs have more than 20 students. Response 3 in figure 8.3 and the sample associated with a minimal response in figure 8.2 are typical examples of this response pattern. From the written responses, it is not at all clear what led them to conclude that graph B had more than 20 students. Giving this problem to students in a classroom setting and asking them to elaborate on their explanations may provide some additional information on this error.

Based on students' explanations about why graph C was the correct answer, the predominant reason given involved the fact that graph C had more students represented (44 students) and the other graphs had fewer students. Response 4 in figure 8.3 illustrates this misinterpretation. It could be that some fourth-grade students focused only on the "most students" part of the phrase "most students had pockets," and chose the graph that had the greatest number of icons with each icon representing one student.

The Graphs of Pockets problem, then, is an excellent example of a task that could help students learn to read and interpret information from

Response 1

Which of the graphs most likely shows the number of pockets that each child had? __B__

Explain why you chose that graph.

I chose it because most of the students have pockets, and there are 23 students.

Explain why you did not choose the other graphs.

They don't follow the rules.

Response 2

Which of the graphs most likely shows the number of pockets that each child had? __B__

Explain why you chose that graph.

I chose graph B because it has most of the people listed as having pockets.

Explain why you did not choose the other graphs.

I didn't chose graph A because it only has a few people with pockets. I didn't chose graph C because it said that everyone had pockets.

Continued

Response 3

Which of the graphs most likely shows the number of pockets that each child had? ___A___

Explain why you chose that graph.

because it has 2 0 students

Explain why you did not choose the other graphs.

because they had over 20 students

Response 4

Which of the graphs most likely shows the number of pockets that each child had? ___letter c___

Explain why you chose that graph.

Because the third graph had 44 students and that was the most student with pockets.

Explain why you did not choose the other graphs.

Because they had less than 44 children with pocket

Fig. 8.3 Graphs of Pockets: Additional student responses

pictographs. NAEP results and the preliminary analysis done for this chapter of a sample of student responses suggest that fourth-grade students had difficulty attending to the dual constraints and writing complete and clear mathematically based justifications for their choices. As with the other extended constructed-response questions, these types of tasks can reveal a great deal about student thinking. Students need more experience with substantive problems requiring them to explain their reasoning in clear and complete ways.

Line Graphs and Scatterplots

Line graphs and scatterplots are often introduced in the middle school curriculum. They not only are found in everyday visual displays of data but are often used to introduce the study of functions and algebra. For example, when representing trends over time, the line graph can represent a particular function that relates the two variables. For scatterplots, when the two variables have a high correlation, a line of best fit can be estimated, providing a linear function that can be used to make predictions about the placement of additional data points. Because of the complex concepts often represented in line graphs and scatterplots, the eight NAEP items containing these representations of data were administered only to students in grade 8 and grade 12. Table 8.5 contains a summary of performance on those eight items.

Notable trends were that the twelfth-grade students' performance improved over that of the eighth-grade students and that almost three-fourths of the twelfth-grade students were successful on all but one of the items involving line graphs. That item, described as item 5 in the table, was an extended constructed-response item asking students to interpret and reason about data in a graph and requiring an extensive explanation in writing. Although only 28 percent of the students in grade 12 responded with a satisfactory or extended response, only 25 percent responded with completely incorrect, irrelevant, or off-task responses, which indicates that about half the students produced responses that were partially correct. It is likely that the major difficulties with this item lie in the need for writing skills and the ability to create a plausible situation, each of which represents important goals in the NCTM *Curriculum and Evaluation Standards for School Mathematics* (1989), that is, to provide opportunities for students to communicate mathematically and to make connections between data representations and realistic situations.

Performance levels on the items involving scatterplots was quite a bit lower than on those involving line graphs. Only about one-third of the twelfth-grade students answered the scatterplot items correctly. However, each item required additional knowledge and procedures that may have added to the difficulty. For example, one item, described in table 8.5 as item 7, required a knowledge of writing equations for lines. Item 8, a regular constructed-response item, required students to evaluate the accuracy of predictions made from scatterplot data. About half the students in grade 12 could identify the correct prediction, but only 20 percent were able to explain why their choice was correct.

The item and results shown in table 8.6 illustrate that the scatterplot items required additional knowledge and procedures in order to obtain

Table 8.5 Performance on Items Involving Line Graphs and Scatterplots

Item Description	Percent Correct	
	Grade 8	Grade 12
Line Graphs		
1. Given a graph with two intersecting lines, determine the point of intersection. (time vs. distance)	65	75
2. Read and compare data points from a graph. (year vs. cost)	49	72
3. Read and compare data points from a graph. (time vs. temperature)	—	75
4. Select the graph that best represents the relationship between data in a table. (time vs. calories burned)	—	72
5. Interpret and reason about data in a graph. (time vs. speed)	—	28[a]
Scatterplots		
6. Use data points to determine the median. (age vs. frequency of activity)	23	31
7. Select the equation representing the line of best fit. (age vs. earnings)	—	35
8. Evaluate the accuracy of predictions based on data in a scatterplot.	—	20

[a]Percent of students scoring at either the satisfactory or extended levels.

Note: Item 5 was an extended constructed-response question, item 8 was a regular constructed-response item, and the rest were multiple-choice items.

the correct answer. Here, students had to read and interpret each data point and then determine the median of the distribution of points as they are related to the number of sit-ups. Thus, that item combined knowledge of the ways in which data are represented and knowledge of statistical terminology. Only a little more than 20 percent of the eighth-grade students and fewer then one-third of the twelfth-grade students selected the correct answer. The most commonly selected distracter for eighth-grade students was 45 (choice C), perhaps a misreading of the points as they correspond to the y-axis. Another one-fourth of the eighth-grade students and almost one-third of the twelfth-grade students selected 55 (choice E), the distracter

Table 8.6 Scatterplot Item

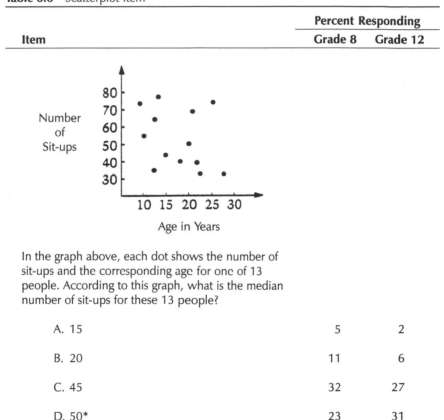

In the graph above, each dot shows the number of sit-ups and the corresponding age for one of 13 people. According to this graph, what is the median number of sit-ups for these 13 people?

Item	Percent Responding	
	Grade 8	Grade 12
A. 15	5	2
B. 20	11	6
C. 45	32	27
D. 50*	23	31
E. 55	26	32

* Indicates correct response

Note: Percents may not add to 100 because of rounding or omissions.

that was consistent with finding the arithmetic mean of the labels given on the vertical axis, that is, adding the labels on the y-axis $(30 + 40 + 50 + 60 + 70 + 80)$ and dividing by 6. This raises questions about students' confusing the meanings of the terms *median* and *mean*, as well as possible misconceptions about how data points are represented in a scatterplot. There is some evidence in NAEP, then, that the difficulty associated with scatterplot items may be compounded by other specialized knowledge that each of these tasks demands.

STATISTICS AND PROBABILITY

Although statistics and probability have traditionally been introduced at the high school and postsecondary levels, the role of these topics is changing in school mathematics. Because topics in probability and statistics across the grades are recommended in the *Curriculum and Evaluation Standards* and because more textbooks and curriculum materials are including those topics, students are becoming acquainted with statistics during their elementary and secondary years of schooling. The 1992 NAEP mathematics assessment included twenty-five items that assessed the topics of statistics and probability. Performance on each topic is discussed in the sections that follow.

Statistics

Eight items addressing concepts in descriptive statistics were administered to students in grades 8 and 12. These items dealt mainly with median and mean as measures of central tendency and with the range as a measure of variability. Some items assessed techniques and procedures such as finding the median or mean for a set of data, and other items addressed concepts of, and comparative meaning for, the different summary statistics. For students in grade 8, performance on these items was very low, with percent-correct values ranging from 15 percent to 38 percent. Performance levels for students in grade 12 were very uneven, with percent-correct values ranging from a high of 72 percent on an item that assessed students' ability to recognize superfluous information to lows of about 20 percent on each of two items that assessed students' understanding of median.

Twelfth-grade students' apparent confusion about the median warrants a closer look. One of the items involved finding the median of a set of numbers, and the other involved evaluating a set of data that included extreme points and identifying the appropriate statistic that best describes a "typical" data point. That twelfth-grade students' performance levels on both median-related items were at the level associated with guessing suggests that those students have neither a well-developed procedural knowledge of how to find the median in a set of data nor an understanding of the concept of median. Further, these results are similar to those associated with twelfth-grade students' performance on the median item in table 8.6 discussed previously. Unfortunately, because no NAEP items assessed students' conceptual understanding of mean and mode for data (except one complex item that involved grouped data in a histogram), there is no way to know whether students have difficulty in general with measures of central tendency or specifically with the median.

Probability

Seventeen items assessed probability concepts, including those on sample space and counting principles, and how to determine the probability of a simple event. Four items were administered only to fourth-grade students, four items only to eighth-grade students, and two items only to twelfth-grade students, with the remaining seven items administered at contiguous grade levels to facilitate performance comparisons. The discussion of student performance on probability concepts focuses on two subsets of items: those associated with identifying a sample space and those dealing with probability as a measure of chance.

Sample Space

Determining the size of the sample space is necessary for analyzing the probability of simple events because the total number of outcomes is part of the ratio representing probability. There were four NAEP items dealing with sample space; one was a multiple-choice item and the other three were regular constructed-response items.

Two of the three regular constructed-response items were secure, and all three items required students to generate a sample space. Those three items are described in table 8.7. The format of a constructed-response item is ideal for assessing students' understanding of sample space as a concept in probability, but the small number of items overall and at each grade level makes it somewhat difficult to generalize patterns in performance. Some notable trends, however, are apparent. In item 1, performance improved from grade 4 to grade 8, with 24 percent of the fourth-grade students and 59 percent of the eighth-grade students responding correctly. The performance of the twelfth-grade students on item 3 was similar to that of eighth-grade students on item 1. However, performance by both eighth- and twelfth-grade students on item 2 was notably lower. Despite the fact that the descriptions given in table 8.7 make the three items appear to be virtually identical, item 2 differs from the other two items in two ways, which may provide possible explanations for the lower level of

Table 8.7 Performance on Sample-Space Items

Item Description	Percent Correct		
	Grade 4	Grade 8	Grade 12
1. List all possible outcomes from picking two colored marbles from a bag.	24	59	—
2. List all possible combinations for a set of objects with replacement.	—	13	24
3. List all possible arrangements of three objects.	—	—	62

Note: All were regular constructed-response items.

performance. First, item 2 was the only one of the three items that did not provide students with an example element in the sample space. This fact suggests that students may have a much easier time once they have been given an illustration of what the response is supposed to look like. The second interesting aspect is that in item 2 the sample space was much larger than that in the other two items, which suggests that the more elements students had to consider or generate, the lower their performance.

The somewhat uneven performance on the NAEP items about sample space is consistent with the stages of development in probability theory as developed by Piaget and Inhelder (1975), as described in Lajoie, Jacobs, and Lavigne (1995). Basing their theory on clinical interviews, Piaget and Inhelder noted that between the ages of 7 and 12, students are in the concrete operational stage. At this stage, students understand the concept of uncertainty, but they must concretely test each case to decide whether or not uncertainty is a possible outcome. Not until the formal operational stage, characterized by abstract thinking, can students generate all possible outcomes based on an internal logical analysis of the situation.

Concept of Probability

The probability of an event is represented by the ratio of the desired outcome(s) to the total number of possible outcomes. The terms *chance* and *likelihood* are often associated with probability. Most of the eight NAEP items summarized in table 8.8 used the word *probability*, but some items administered to fourth-grade students used alternative language and words. However, successful responses in all instances required students to interpret the ratio of desired outcome(s) to possible outcomes.

Results for items 1, 2, and 3 in table 8.8 show that more than half of fourth-grade students chose correct answers to multiple-choice items that required identifying one favorable outcome out of a given number of out-comes. However, the performance for students at that grade level on items 4 and 6 was substantially lower—at a level not much better than guessing. One possible reason for students' lower performance levels on those two items involves the fact that in both items students had to find the total number of possible outcomes, whereas in items 1, 2, and 3 the total number of outcomes was given information. Because only one probability item was administered to students in grade 12, little can be said about performance at that grade level. Because of the small number of probability items administered to eighth-grade students, it is somewhat difficult to interpret the results. Although eighth-grade students generally did well on the multiple-choice items (items 5 and 6), with percent-correct values ranging from 55 percent to 73 percent, once again performance levels dropped on the two regular constructed-response items (items 7 and 8).

The disparity in performance between probability items presented in multiple-choice format and those presented in constructed-response

Table 8.8 Performance on Probability Items

Item Description	Percent Correct		
	Grade 4	Grade 8	Grade 12
1. Find simple probability (1 out of N).	59	—	—
2. Find simple probability (1 out of N).	50	—	—
3. Identify the situation with the greatest chance of occuring (Each situation: 1 out of N).	53	—	—
4. Find the probability (x out of N).	26	—	—
5. Given the proportions of objects in a bag, select the most likely outcome.	25	73	—
6. Find the probability (x out of N).	23	55	—
7. Find the probability and explain (x out of N).	—	13[a]	—
8. Find the probability (x out of N).	—	18	29

[a]Percent of students scoring at either the satisfactory or extended levels.

Note: Item 7 was an extended constructed-response question, item 8 was a regular constructed-response item, and the rest were multiple-choice items.

format may be explained in part by research done by Green (1983). On the basis of tests administered to about 3,000 students and on follow-up interviews of forty-four of those students, Green found that students between the ages of 11 and 16 could often provide correct answers to probability questions without any understanding of the underlying concepts and mathematical processes of probability. In the case of NAEP items, then, students who can respond correctly to multiple-choice items may not have the underlying knowledge about probability that would enable them to explain their reasoning.

One item, shown in table 8.9, required students to use their understanding of fractions to deal with a probability situation. There was great disparity in performance between fourth- and eighth-grade students. An examination of the performance across the distracters reveals that fourth-grade students especially were drawn to the one associated with the fraction with the greatest denominator. In fact, 56 percent of those students chose this response compared with only 18 percent of eighth-grade students, which perhaps indicates a growth in understanding of fractions over time. Other probability items in NAEP also point to the difficulty that fractions may pose for students. For example, on an item that required students to interpret a probability in fraction form and then use it to determine the expected value, only 37 percent of eighth-grade students responded correctly.

Table 8.9   Probability Item Using Fractions

Item	Percent Responding	
	Grade 4	Grade 8
In a bag of marbles, $\frac{1}{2}$ are red, $\frac{1}{4}$ are blue, $\frac{1}{6}$ are green, and $\frac{1}{12}$ are yellow. If a marble is taken from the bag without looking, it is most likely to be		
A. red*	25	73
B. blue	7	4
C. green	12	4
D. yellow	56	18

*Indicates correct response.
Note: Percents may not add to 100 because of rounding or omissions.

OVERALL PERFORMANCE ON DATA ANALYSIS, STATISTICS, AND PROBABILITY

Reports of NAEP results included looking at performance across the entire set of data analysis, statistics, and probability items according to the NAEP scale. Table 8.10 contains data on the average performance on this set of items for students at each grade level and for selected subgroups of students. Because no performance scale was created in 1990 for grade 4, the data for 1990 are restricted to grades 8 and 12 only. For those two grades levels, performance increased significantly from 1990 to 1992, and in 1992

Table 8.10 Average Performance in Data Analysis, Statistics, and Probability: Overall Performance and Performance by Gender and Race/Ethnicity, Grades 4, 8, and 12 for 1990 and 1992

	Grade 4	Grade 8	Grade 12
Overall			
1992	219 (0.9)	268 (1.1)>	298 (1.0)>
1990	—	263 (1.6)	294 (1.2)
Gender			
Male			
1992	220 (0.9)	268 (1.3)>	299 (1.2)
1990	—	264 (1.9)	297 (1.4)
Female			
1992	219 (1.2)	269 (1.2)>	296 (1.1)>
1990	—	263 (1.6)	292 (1.5)
Race/Ethnicity			
White			
1992	228 (1.1)	279 (1.2)>	305 (1.0)
1990	—	273 (1.6)	302 (1.3)
Black			
1992	192 (1.6)	235 (1.7)	273 (1.9)>
1990	—	233 (3.2)	266 (2.3)
Hispanic			
1992	203 (1.4)	243 (1.5)	281 (2.2)
1990	—	240 (3.3)	275 (3.7)

Note: At grade 4 no trend data are available for this content category. The fifth NAEP mathematics assessment (1990) at grade 4 contained too few items to create a performance scale.

> The value for 1992 was significantly higher than the value for 1990 at about the 95 percent confidence level. The standard errors of the estimated proficiencies appear in parentheses.

performance increased steadily between contiguous grade levels, especially from grade 4 to grade 8. Data from the NAEP teacher questionnaires for those two grades reveal that emphasis on concepts in this content area increases from the elementary grades through middle school. In particular, although fewer than half the fourth-grade students had teachers who reported giving heavy or moderate emphasis to data analysis, statistics, and probability, by grade 8 more than 70 percent of those students had teachers who emphasized this content area.

Performance by both males and females at grade 8 increased from 1990 to 1992, but only females at grade 12 had a significant performance increase. Comparing males and females within a grade level, performance levels in data analysis, statistics, and probability were nearly identical for all three grade levels and nearly identical to the overall performance levels. For race/ethnicity subgroups, White students at grade 8 and Black students at grade 12 had significantly higher performance levels in 1992 than in 1990. Among the subgroups in 1992, White students outperformed both Black and Hispanic students at all three grade levels, and Hispanic students outperformed Black students at grades 4 and 8.

CONCLUSION

The 1992 NAEP results point to conclusions about students' performance on data analysis, statistics, and probability. About half the fourth-grade students and most of the eighth- and twelfth-grade students were able to read, use, and complete data entries in and from tables, pictographs, and bar graphs. About three-fourths of the twelfth-grade students were able to read and use data from line graphs. There was confusion about the meaning of the measures of central tendency, especially that of median, for eighth- and twelfth-grade students. Finally, when probability involved a single favorable outcome, more than half the fourth-grade students were able to find a ratio for the probability and to identify a situation consistent with a given probability.

Of particular note in the 1992 assessment was the opportunity to examine students' responses on regular and extended constructed-response items. The addition of these types of tasks to the assessment instrument provides information about communication, reasoning, and problem solving, in particular, the difficulty students have in explaining their thinking in writing and communicating about important mathematical ideas. For example, as shown in table 8.5, performance on multiple-choice items that involved reading and interpreting data from line graphs ranged from 72 percent to 75 percent correct for twelfth-grade students, but successful performance dropped to 28 percent on an extended constructed-response item that involved an extensive written explanation in addition to the interpretation of a line graph. These discrepancies in performance levels raise such questions as the following: What does it really mean to be able to interpret a line graph? and What different types of knowledge about line graphs are required for these two different types of items?

The inclusion of regular and extended constructed-response items on the NAEP mathematics assessment affords opportunities to raise such questions and to explore very different kinds of implications for curriculum and instruction. Another benefit of including these types of items is that although in-depth analyses of incorrect open responses are not formally included in the NAEP reporting scheme, the availability of the student responses provides a way to access more information about the nature of students' understandings, to examine their ability (or inability) to communicate mathematically, and to gain insights about their reasoning. For example, the examination of student responses to the Graphs of Pockets problem and the Trash Cans problem described in this chapter yielded qualitative information that can be useful in thinking about further research and development in the curriculum of data analysis, statistics, and probability. Overall, the information from the combination of multiple-choice and constructed-response items for graphical representations painted a richer, more informative picture of students' ability to read and interpret data.

REFERENCES

Dossey, John A., Ina V. S. Mullis, and Chancey O. Jones. *Can Students Do Mathematical Problem Solving? Results from Constructed-Response Questions in NAEP's 1992 Mathematics Assessment*. Washington, D.C.: National Center for Education Statistics, 1993.

Green, David R. "A Survey of Probability Concepts in 3000 Pupils Aged 11–16 Years." In *Proceedings of the First International Conference on Teaching Statistics*, edited by D. R. Grey, P. Holms, V. Barnett, and G. M. Constable, pp. 766–83. Sheffield, England: Organising Committee of the First International Conference on Teaching Statistics, 1983.

Lajoie, Susanne P., Victoria R. Jacobs, and Nancy C. Lavigne. "Empowering Children in the Use of Statistics." *Journal of Mathematical Behavior* 14 (December 1995): 401–25.

National Assessment of Educational Progress. *Mathematics Objectives: 1990 Assessment*. Princeton, N.J.: Educational Testing Service, National Assessment of Educational Progress, 1988.

National Council of Teachers of Mathematics. *Curriculum and Evaluation Standards for School Mathematics*. Reston, Va.: National Council of Teachers of Mathematics, 1989.

Piaget, Jean, and Barbel Inhelder. *The Original Idea of Chance in Children*. Translated by Lowell Leake, Jr., Paul Burrell, and Harold D. Fishbein. New York: W. W. Norton, 1975.

9

What Do Students Know about Algebra and Functions?

Glendon W. Blume & David S. Heckman

ALGEBRA AND FUNCTIONS are receiving increased emphasis in the school mathematics curriculum. This is evidenced by increased attention in the elementary mathematics curriculum to informal development of the ideas of algebra, including recognizing, describing, extending, and creating patterns; representing and describing mathematical relationships; and using variables and number sentences to express relationships. At the secondary level, the importance of algebra and functions is evidenced by an increased curricular emphasis on patterns, the concept of function, and function notation; a decreased emphasis on by-hand graphing because of the use of graphing calculators; and the increased inclusion of algebra as a requirement for graduation from high school.

The purpose of this chapter is to summarize what students know about algebra and functions on the basis of results from the 1992 NAEP mathematics assessment, ranging from informal knowledge of algebraic ideas to more-advanced topics such as algebraic proofs and trigonometric functions. This chapter is intended to give insights into student performance that will establish a base from which teachers and researchers can begin to address teaching and learning issues that arise when students encounter graphic, numeric, and symbolic representations of functions; when graphing calculators become more widely available to students; when computer algebra systems, such as the Texas Instruments TI-92, become available to students; and when schools begin to implement algebra for all students.

ALGEBRA AND FUNCTIONS IN THE
NAEP FRAMEWORK AND ITEMS

In the NAEP framework, the content area of algebra and functions was defined to be "broad in scope, covering a significant portion of the grade

9–12 curriculum, including algebra, elementary functions (precalculus), trigonometry, and some topics from discrete mathematics" (National Assessment of Educational Progress 1988, pp. 28–29). The framework document emphasized that proficiency in this content area required the ability to use algebra as a means of representation and algebraic processes in problem solving, and viewed functions not only in terms of symbolic function rules or formulas but also in terms of verbal descriptions, tables of values, and graphical representations. At the elementary and middle school grades, algebra and functions topics were to be treated informally and in exploratory ways. The framework recommended the following distribution of NAEP questions for algebra and functions: 10 percent at grade 4, 20 percent at grade 8, and 25 percent at grade 12. Specific topics deemed appropriate by NAEP for fourth-grade students involved describing, extending, and creating a variety of patterns and functional relationships; relating symbolic expressions and verbal statements; using number lines and rectangular coordinate systems; and solving simple linear equations and inequalities

HIGHLIGHTS

- Students at all grade levels generally could recognize number patterns presented in tables.
- Finding a relationship between two quantities was more difficult than finding terms that would continue a pattern. Patterns with a nonconstant rate of increase were more difficult for younger students than patterns with constant increases or decreases.
- By grade 8 more than half of the students were successful on items that embodied informal treatments of algebraic ideas.
- Although performance improved from grade 8 to grade 12, the majority of students had difficulty solving equations and inequalities other than fairly simple ones.
- About half of twelfth-grade students could correctly answer items that use function notation, but they had more difficulty drawing conclusions about graphs of functions or identifying particular functions from those graphs.
- In general, on extended constructed-response questions that required the justification of results and communication in writing, only a very small percentage of students were successful. Doing and communicating about mathematics was more difficult for students than merely doing mathematics.
- Performance levels in algebra and functions increased significantly from 1990 to 1992 for students in all three grades, with a sharper increase between grades 4 and 8 than between grades 8 and 12.

using whole numbers. The topic list at grade 8 was expanded to include working with algebraic expressions and equations in monomial, polynomial, or rational form involving one or more variables and further expanded at grade 12 to include discrete mathematics and trigonometric concepts.

The 1992 NAEP mathematics assessment contained seventy-two algebra and functions items, with thirty-five items administered only to twelfth-grade students, twenty items administered to fourth- or eighth-grade students, and seventeen items administered at more than one grade level to facilitate comparisons of student performance between or among grade levels. About two-thirds of the items were in multiple-choice format, and the remaining one-third were constructed-response items that required students to produce their own answers or explanations. Five items were extended constructed-response questions that required students to explain their answers and reasoning processes in writing. While working on certain algebra and functions items, fourth-grade students were permitted to use simple, four-function calculators, and eighth- and twelfth-grade students were permitted to use scientific calculators. The sections that follow discuss student performance on individual NAEP algebra and functions items and clusters of related items in these categories: patterns, relationships, and informal algebra; graphing on the number line and in the coordinate plane; simplifying and evaluating expressions; solving and graphing equations and inequalities; systems of linear equations; evaluating and graphing functions; algebraic reasoning; and trigonometric functions.

PATTERNS, RELATIONSHIPS, AND INFORMAL ALGEBRA

Mathematics often is viewed as the study of patterns. In fact, Steen (1990) notes that "mathematics is not just about number and shape but about pattern and order of all sorts" (p. 2). The NCTM *Curriculum and Evaluation Standards for School Mathematics* (NCTM 1989) stresses the importance of recognizing, describing, extending, and creating patterns. Because such processes provide experiences in working with relationships between variables, they are important to the development of the concept of function. In the early grades, "the idea of a functional relationship can be intuitively developed through observations of regularity and work with generalizable patterns" (NCTM 1989, p. 60). By describing and representing relationships with tables, graphs, and rules as advocated in the *Standards* (NCTM 1989), students in grades 5–8 can progress from viewing patterns as predictable changes in the values of one variable to viewing patterns as dynamic relationships between two variables. As students begin to develop their capabilities to generalize about those relationships, they also develop an informal understanding of function that prepares them to work more formally with numeric, graphic, and symbolic representations of functions in algebra. Students' experiences with patterns provide a bridge from numeric work at the elementary level to more general, symbolic algebra at the secondary level.

The structure of the NAEP content area of algebra and functions stipulated that algebraic and functional concepts be treated in more informal, exploratory ways in grade 4 and grade 8. The majority of items administered to students at those grades presented patterns numerically and addressed functions informally, without requiring students to use general symbolic descriptions or function notation. The twelfth-grade items involved more-explicit references to functions and their graphs and symbolic descriptions. The following sections examine results related to patterns, relationships, and informal algebra.

Patterns and Relationships

The NAEP items related to patterns can be separated into two categories: those involving a *repeating* pattern and those involving a *growing* pattern. A repeating pattern item involves elements that in some way remain the same, for example, a single symbol presented in a sequence of different positions. In contrast, a growing pattern involves elements that change in some predictable way, for example, a sequence of numbers that increases (or decreases) by a fixed amount.

Repeating Patterns

Two NAEP items, each administered across grade levels, involved repeating patterns. The elements in these patterns were nonnumeric, that is, geometric figures and alphabetic characters. The first item was in multiple-choice format and displayed four orientations of a figure: ⊔ ⊓ ⊏ ⊐. Students were asked to determine which of the four given orientations would be next in the pattern ⊔⊏⊓⊐⊔⊏⊓⊐⊔__. While the designated correct response was the symbol opening to the right, it is important to realize that there is an inherent ambiguity in items such as these. For a question such as "Which figure would be next?" an infinite number of correct responses exist because an infinite number of patterns could begin with the given nine elements. For example, the upward opening symbol might be the correct choice for the next figure if the pattern involves repetition of the first nine symbols in blocks of nine. Nevertheless, 91 percent of students in grade 4 and 96 percent in grade 8 chose the designated correct response that was consistent with the repetition of blocks of four symbols that appeared in the item.

Students had much more difficulty with the other repeating pattern item, a regular constructed-response item involving completion of missing terms in a sequence of letters. Three factors may have contributed to the difficulty of this item: its constructed-response format, students' being asked to supply elements at various positions in the pattern rather than just the "next" element, and fewer given elements from which to discern the pattern. Although only about one-third of fourth-grade students answered that item correctly, more than half the eighth-grade students and two-thirds of twelfth-grade students correctly provided the missing letters. There was more improvement from grade 8 to grade 12 than from grade 4 to grade 8, perhaps because work with patterns and sequences in the early grades focuses initially on finding the next element in a pattern, whereas work with sequences at the high school level includes having students find preceding terms when subsequent terms of a sequence are given.

Growing Patterns

Twelve algebra and functions items addressed growing patterns, three of which asked for the next number(s) in the pattern and nine of which asked for numbers other than the one(s) that came immediately after those that were given. Almost all of those items were administered to students in grade 4 or grade 8, with only one administered to twelfth-grade students. A summary of the growing pattern items and the results at grades 4 and 8 appears in table 9.1. On items that involved finding the next number in the pattern, approximately one-third to one-half of fourth-grade students responded correctly. Performance was lower when students in grade 4 were asked to find numbers other than one(s) that came next in the pattern, with percent correct in many cases ranging from 20 percent to 33 percent. Eighth-grade students also had difficulty with more-complex pattern questions.

The percent-correct values in table 9.1 for the items that required fourth-grade students to find the next numbers in patterns show that performance was somewhat uneven among the three items, that is, performance was lower on item 3 in the table (32 percent) than on the other two items (55 percent and 51 percent). A closer look at two of those items, shown in table 9.2, reveals some interesting information about those items that may have had an effect on performance.

Item 1 in table 9.2 asked students to use an explicitly stated constant increase of 50 students per year to find the number of students expected in each of three years. While working on that item, students had the use of a simple, four-function calculator. About half the fourth-grade students correctly found the next three numbers in item 1, and, in fact, including the additional 12 percent of students who found either one correct entry or two correct entries, the performance increases to more than 60 percent of students who are at least partially successful in finding the next terms in a growing pattern. With respect to whether using a calculator had any effect on performance, results show that for students who reported using the calculator, 59 percent obtained all three correct values for the number of

Table 9.1 Performance on Pattern Items

Item Description	Percent Correct	
	Grade 4	Grade 8
Identifying the next number(s) in a pattern		
1. Choose the correct sum for several numbers in a pattern.	55	—
2. Given an increase of 50, find the next three numbers in a table.	51	—
3. Choose the correct value based on the pattern of increases in a puppy's weight.	32	66
Identifying other number(s) in a pattern		
4. Choose a number that would be a term in a given pattern.	29	—
5. Reason about whether or not a given number could be in a pattern and explain.	27	—
6. Use a rule to generate another number in a number pattern in a table.	23	—
7. Apply pattern recognition to choose the number of tacks needed to hang pictures.	25	48
8. Extend a pattern in a geometric situation.	—	50
9. Extend a pattern and choose the correct value.	—	33
10. Extend a pattern in a table and choose the correct value.	—	25
11. Determine the 20th element in a pattern and explain how to find it.	—	6[a]

[a]Percent of students scoring at either the satisfactory or extended levels.

Note: Item 11 was an extended constructed-response question; items 2, 5, 6, and 8 were regular constructed-response items; and the rest were multiple-choice items.

students as opposed to 49 percent correct for those students who said they did not use the calculator. The difference between the two percent-correct values is significant at the .05 level.

For item 2, which was administered to both fourth- and eighth-grade students and which involved a decreasing rate of increase that was not explicitly revealed to students, only about one-third of fourth-grade students chose the correct answer. The two distracters selected most often were 25 pounds (29 percent) and 27 pounds (24 percent). Whereas some students may have just guessed, other students may have selected 25 because

Table 9.2 Growing Pattern Items: Next Number(s)

Item	Percent Responding	
	Grade 4	Grade 8

1ª. In 1990 a school had 125 students. Each year the number of students in the school increases by 50. Fill in the table to show the number of students expected for each year.

Year	Number of Students
1990	125
1991	—
1992	—
1993	—

	Grade 4	Grade 8
All three answers correct (175, 225, 275)	51	—
One or two correct answers	12	—
Incorrect or incomplete responses for all three years	30	—
Omitted	7	—

Puppy's Age	Puppy's Weight
1 month	10 lbs.
2 months	15 lbs.
3 months	19 lbs.
4 months	22 lbs.
5 months	?

2. John records the weight of his puppy every month in a chart like the one shown above. If the pattern of the puppy's weight gain continues, how many pounds will the puppy weigh at 5 months?

	Grade 4	Grade 8
A. 30	12	4
B. 27	24	14
C. 25	29	14
D. 24*	32	66

*Indicates correct response.

ªItem is from an item block for which students were provided with, and permitted to use, simple, four-function calculators.

Note: Percents may not add to 100 because of rounding or omissions.

they assumed the rate of increase remained constant, so the increase of 3 pounds from 3 months to 4 months was used as the increase from 4 months to 5 months. Of the students who selected 27, some could have assumed a constant rate of increase of 5 (the increase from 1 month to 2 months) or they merely could have added 5 months and 22 pounds to get the next number, 27, in the pattern. The encouraging news is that 66 percent of the eighth-grade students answered correctly, which suggests that errors involved with assuming a constant rate of increase rather than the given nonconstant rate of increase were not as prevalent in grade 8.

Performance on those two items suggests that patterns based on nonconstant rates of change are more difficult for fourth-grade students than patterns based on constant rates of change. This finding is consistent with the difficulty secondary students have when working with nonlinear functions such as quadratic and exponential functions and with students' inclinations to view nonlinear situations as linear (Markovits, Eylon, and Bruckheimer 1983). Because some students in grades 4 and 8 may expect number patterns always to entail a constant increase or decrease, this points to the need to provide elementary and middle school students with experiences in which they complete or extend, reason about, and identify the underlying rules for, a variety of patterns that embody both constant and nonconstant rates of change.

Another set of growing pattern items, described in table 9.1 as items 4–11, required students to find numbers other than those that immediately followed ones given in the item or required students to reason about numbers that appeared later in the pattern. In general, these growing pattern items were more difficult for students than ones that asked only for the next term or terms. A closer examination of responses to some of the growing pattern items that involve terms further into the pattern gives some insight into student performance.

Items 1 and 2 in table 9.3 were an item pair that involved identifying the rule used to create the pattern of numbers in the table and then using that rule to produce another table entry. In the first item, 42 percent of fourth-grade students chose the correct response of choice A, "Divide the number in column A by 4," and 26 percent selected choice B, "Multiply the number in column A by 4"—the correct magnitude of the multiplicative relationship but the wrong operation. It is encouraging that nearly 70 percent of fourth-grade students identified the multiplicative relationship, while only about 25 percent chose the responses that reflected subtractive or additive relationships, that is, subtraction of 9 in choice C or addition of 9 in choice D. Item 2, a regular constructed-response item that asked students to use the rule they found in item 1 to find the number that corresponded to 120, was more difficult for students than the multiple-choice item in the pair. Twenty-three percent wrote the correct answer and 18 percent omitted

Table 9.3 Growing Pattern Items: Number(s) Other than the Next One(s)

Item	Percent Responding	
	Grade 4	Grade 8

Column A	Column B
12 →	3
16 →	4
24 →	6
40 →	10

1ª. What is a rule used in the table to get the numbers in column B from the numbers in in column A?

	Grade 4	Grade 8
A. Divide the number in column A by 4.*	42	—
B. Multiply the number in column A by 4.	26	—
C. Subtract 9 from the number in column A.	17	—
D. Add 9 to the number in column A.	8	—

2ª. Suppose 120 is a number in column A of the table. Use the same rule to fill in the number in column B.

Column A	Column B
120 →	

	Grade 4	Grade 8
Correct response of 30	23	—
Incorrect response of 480	4	—
Incorrect response of 111	5	—
Incorrect response of 129	1	—
Any other incorrect response	49	—
Omitted	18	—

Continued

3. Children's pictures are to be hung in a line as shown in the figure above. Pictures that are hung next to each other share a tack. How many tacks are needed to hang 28 pictures in this way?

	Grade 4	Grade 8
A. 27	27	18
B. 28	27	21
C. 29*	25	48
D. 56	19	11

A	B
2	5
4	9
6	13
8	17
14	?

4. If the pattern shown in the table were continued, what number would appear in the box at the bottom of column B next to 14?

	Grade 4	Grade 8
A. 19	—	4
B. 21	—	45
C. 23	—	7
D. 25	—	17
E. 29*	—	25

*Indicates correct response.
ªItem is from an item block for which students were provided with, and permitted to use, simple, four-function calculators.
Note: Percents may not add to 100 because of rounding or omissions.

item 2, even though only 6 percent omitted the first item of the pair. Only 4 percent responded "480," suggesting that few students multiplied, even though about one-fourth of the students chose multiplication by 4 on the previous item. The fact that the number 120 was a much larger number than the numbers shown in the table may have suggested to some students that they should divide rather than multiply. Also, only 5 percent responded "111," a value that could be obtained by subtracting the difference of 12 and 3 in the first pair from 120.

This item pair appeared in a calculator block of items and was classified as calculator neutral. On item 1, there was little difference in performance between those who reported they used the calculator and those who reported they did not, with percent-correct values of 48 percent and 45 percent, respectively. However, on item 2, 42 percent of those who reported using the calculator answered correctly, whereas only 12 percent of those who reported that they did not use the calculator answered correctly. That difference is significant at the .05 level, but one needs to be careful not to impute improved performance to calculator use. It may be that the number 120 was large enough to warrant calculator use, and students who scored well may have been the ones who typically chose to use the calculator.

Students were not consistent in their answers across the two items—only about half of the students who selected the correct rule in the first item applied that rule correctly in the second item. Unfortunately, the existing NAEP data analyses provide no information on the types of errors students made in implementing the rule they correctly chose. Performance on this item pair does, however, suggest that many fourth-grade students can identify aspects of a functional relationship between two quantities even though some of them may not be able to apply the rules they identify to subsequent values in a table.

Another growing pattern item that required students to reason about a pattern appears in table 9.3 as item 3. Although performance by students in grade 4 on this multiple-choice item was at the chance level for a four-choice item, of the 19 percent who selected choice D (56), some students may have reasoned appropriately about the pattern. Students could have interpreted "share a tack" as allowing an additional tack to be used, but hidden, on the paper underneath the overlap. Eighth-grade students were more successful on this item, with nearly half answering correctly. This item, without the answer choices given, could be an interesting one to present to elementary or middle school students. Students might model a portion of the pattern at a bulletin board, and one could observe their work and have them describe their reasoning about the pattern. Potential ambiguity concerning hidden tacks might be diminished by presenting a variation of this problem, for example, a problem set in the context of sections of fence that share fence posts.

Eighth-grade students' performance was mixed on growing pattern items that asked for numbers further removed from the given elements of the pattern. Most percent-correct values ranged from about 20 percent to 50 percent on the multiple-choice and regular constructed-response items, but for an extended constructed-response item (item 11 in table 9.1), fewer than 10 percent of the students produced a satisfactory or extended response. Although performance on some eighth-grade items was higher than performance on fourth-grade items that involved finding numbers further into a pattern, eighth-grade students had difficulty with more-complex patterns.

Item 4 in table 9.3 is an example of a difficult item for eighth-grade students. The two most commonly chosen distracters for this item were 21, the next number in the pattern, and 25, the one that followed it. Thus, nearly two-thirds of the students chose numbers that were consistent with a strategy of simply producing additional terms in the table. Worth noting here is the fact that the percent of eighth-grade students who chose the incorrect answer of 21 was much higher than the percent who chose the correct answer of 29. Some students who were unable to find a functional relationship between the numbers in column A and those in column B—that is, that the number in column B is one more than double the corresponding number in column A—may have been able to recognize the constant increase of 4 in column B. Among students who may have used this constant increase, those who answered correctly simply may have generated the next three entries in the table, while those who were incorrect may have generated too few entries or may not have understood the "break" in the table.

Problem Solving with Patterns

Several pattern items, two of which are described next, asked students to reason about patterns. The first of these items, called Extend Pattern in a NAEP report (Dossey, Mullis, and Jones 1993, pp. 70–73) is shown in table 9.4. This regular constructed-response item asked students in grade 4 to reason about a pattern of products, each being a power of 2, to determine whether 375 would appear in the pattern and to explain why it would or would not appear. While working on this item, students were provided with, and permitted to use, simple, four-function calculators. Responses were scored according to a right-or-wrong scoring scheme.

Although 33 percent of fourth-grade students did not reach this item (it was the last item in the item block), of those who completed it, the majority (76 percent) reported that 375 could not be one of the products in the pattern. However, only 27 percent gave an appropriate explanation for their choice. This suggests that even though young children can generate appropriate answers, communicating their reasoning that led to those answers is often much more difficult. One can gain further insights into students' reasoning on this item by examining some of their explanations.

Table 9.4 Extend Pattern

Item[a]	Percent Responding Grade 4
Product	
$2 \times 2 = 4$	
$2 \times 2 \times 2 = 8$	
$2 \times 2 \times 2 \times 2 = 16$	
$2 \times 2 \times 2 \times 2 \times 2 = 32$	
If the pattern shown continues, could 375 be one of the products in this pattern?	
○ Yes ○ No	
Explain why or why not.	
Correct explanation	27
Any incorrect or incomplete explanation	60
Omitted	13

[a]Item is from an item block for which students were provided with, and permitted to use, simple, four-function calculators.

In a sample of almost 250 complete student responses containing both a Yes or No choice about whether 375 could be in the pattern of products and an explanation that did not contradict that choice, only about thirty respondents chose Yes, with the vast majority choosing No. More than half the Yes responses were based on the notion that if the pattern of numbers continues, the number 375 will eventually appear (for example, "The pattern would have to go a long way, but yes 375 could be in"). Responses 1 and 2 in figure 9.1 illustrate this kind of reasoning. These students may have based their reasoning not on the characteristics of the numbers in the pattern but on the fact that the pattern continued indefinitely. A misunderstanding of infinity may have led them to think that any number would eventually appear in an infinite sequence of products.

The majority of the No responses gave explanations based on number theory concepts involving even and odd numbers. Responses 3 and 4 in figure 9.1 are examples of that kind of explanation. Not all explanations that used odd and even numbers were as well developed, for example, "Because it's a [sic] odd number." A few responses, of which responses 5 and 6 in the figure are examples, mentioned other number theory concepts such as factors and multiples and divisibility. While no responses from students in grade 4 directly mentioned the fact that the products were the

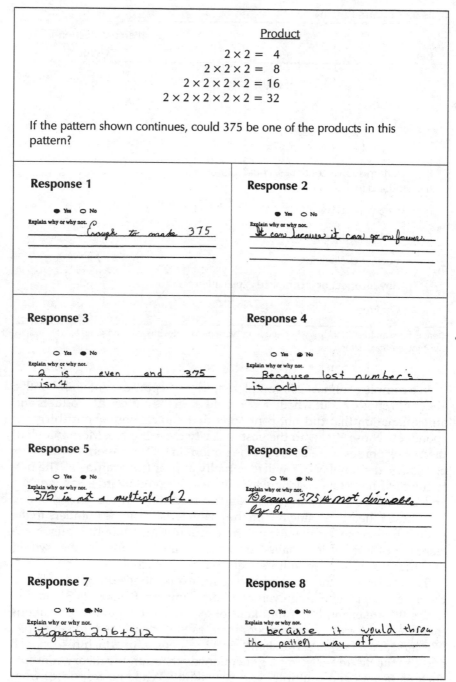

<u>Product</u>

$$2 \times 2 = 4$$
$$2 \times 2 \times 2 = 8$$
$$2 \times 2 \times 2 \times 2 = 16$$
$$2 \times 2 \times 2 \times 2 \times 2 = 32$$

If the pattern shown continues, could 375 be one of the products in this pattern?

Response 1

● Yes ○ No

Explain why or why not.

_____ Enough to make 375

Response 2

● Yes ○ No

Explain why or why not.

It can because it can go on forever.

Response 3

○ Yes ● No

Explain why or why not.

2 is even and 375 isn't

Response 4

○ Yes ● No

Explain why or why not.

Because last number's is odd

Response 5

○ Yes ● No

Explain why or why not.

375 is not a multiple of 2.

Response 6

○ Yes ● No

Explain why or why not.

Because 375 is not divisable by 2.

Response 7

○ Yes ● No

Explain why or why not.

it goes to 256+512

Response 8

○ Yes ● No

Explain why or why not.

because it would throw the patten way off

Fig. 9.1 Extend Pattern: Sample responses

powers of 2, a few responses contained implied references to powers of 2, either by explanations that appeared to have been based on reasoning about powers of 2 (for example, "Because 2 times as many 2s as you want does not add up to 375") or reporting, in essence, that the numbers in the pattern are either greater than or less than 375, as shown in response 7 in the figure.

Some of the reasons given for choosing No were not expressed clearly or completely. For example, the response labeled response 8 in figure 9.1 ("Because it [375] would throw the pattern way off") and others like it suggest that some students in grade 4 intuitively believed that 375 would not be in the pattern of products but could not express their reasoning in mathematical terms.

As mentioned above, the Extend Pattern problem appeared in a block of items for which students were permitted to use calculators, but the item itself was classified by NAEP as calculator neutral, that is, calculator use was considered to be optional with respect to solving the problem. When the kinds of explanations that were identified in the sample of 250 responses were compared to the answer to the calculator-use question, the results showed that most students whose explanations were based on odd and even numbers reported not using the calculator. Similarly, most of those who used divisibility arguments reported no calculator use. Those students who used factors and multiples were approximately equally split between calculator use and no calculator use, as were those who used powers of 2. However, among those whose explanations were based on 375 being between two powers of 2—256 and 512—all reported using the calculator. It appears that students who drew conclusions that were not necessarily based on doing computations appropriately chose not to use the calculator; however, availability of the calculator may have enabled at least some students to use an argument (in particular that 256 and 512 were in the pattern and 375 was not) that might not have been available to them had they not had a tool with which to compute the larger powers of 2.

The Extend Pattern problem illustrates the way in which patterns can be connected to other topics in mathematics, in this case, to number theory concepts. The richness of some students' explanations and the incompleteness of those produced by other students highlight the need for teachers and researchers to examine children's informal algebraic thinking in detail and to provide opportunities in which students can communicate and explain their thinking to others.

Another NAEP item that involved problem solving about patterns was an extended constructed-response question called Marcy's Dot Pattern and shown here in table 9.5. This question, administered to students in grade 8, asked them to find the 20th step in a pattern and provide an explanation of how to do so without finding each of the intervening pictures in

Table 9.5 Marcy's Dot Pattern

Question[a]	Percent Responding Grade 8

[General directions]

This question requires you to show your work and explain your reasoning. You may use drawings, words, and numbers in your explanation. Your answer must be clear enough so that another person could read it and understand your thinking. It is important that you show <u>all</u> your work.

A pattern of dots is shown below. At each step, more dots are added to the pattern. The number of dots added at each step is more than the number added in the previous step. The pattern continues infinitely.

(1st step)   (2nd step)   (3rd step)

Marcy has to determine the number of dots in the 20th step, but she does not want to draw all 20 pictures and then count the dots.

Explain or show how she could do this <u>and</u> give the answer that Marcy should get for the number of dots.

Extended response	5
Satisfactory response	1
Partial response	6
Minimal response	10
Incorrect	63
Omitted	16

[a]Question is from an item block for which students were provided with, and permitted to use, scientific calculators.

Note: Percents may not add to 100 because of rounding.

the pattern of dots. Although students were not expected to generalize in symbolic form—for example, that the nth step would contain $n(n + 1)$ dots—for their answer to be correct, they did need to find the number of dots in the 20th step and relate that number to the earlier given values in some general way. If students have no immediately available means of determining the number of dots in that particular step, this type of generalizing to find a particular element in the pattern without exhaustively creating all intervening elements involves problem solving. For example, students could focus on the relationship between the steps and the number of dots in the rows and columns: one row, two columns, and two (1×2) dots in step 1; two rows, three columns, and six (2×3) dots in the second step; three rows, four columns, and twelve (3×4) dots in the third step; and so on. Continuing the pattern results in twenty rows, twenty-one columns, and 420 (20×21) dots in the 20th step. Marcy's Dot Pattern, then, represents the sort of task that is transitional, from purely informal algebraic tasks that focus on recognizing and applying patterns to more formal algebraic tasks in which general descriptions of functional relationships, often in symbolic form, are the focus.

Responses to Marcy's Dot Pattern were scored according to five performance categories—incorrect, minimal, partial, satisfactory, and extended—and the designation "no response" for blank papers (also called "omitted"). Figure 9.2 contains descriptions of the criteria for each performance category and an illustrative example of a response from each category. The descriptions and examples are as they appear in a NAEP publication (Dossey, Mullis, and Jones 1993, pp. 134–36). Results for this question are shown in table 9.5 and reveal that eighth-grade students did poorly on this question. The vast majority of the students omitted it or wrote responses that were judged to be incorrect. Only about 10 percent of students scored at the "minimal" level, and another 6 percent wrote "partial" responses. Finally, only 6 percent of the eighth-grade students produced either a satisfactory response or an extended response.

Because the five examples in figure 9.2 were selected to represent the performance categories for Marcy's Dot Pattern, they give limited information about how students attempted to answer this extended question. Other responses from a set representing 20 percent of the total number of nonblank responses gathered and analyzed for this chapter afford some additional information. Given the overall poor performance on this question, it was not surprising that in the sample gathered for this chapter there were only a few good examples of responses that successfully explained how Marcy could find the number of dots in the 20th step. Almost half the responses in the sample were incorrect, and most of these incorrect responses were based only on the number of dots shown in the first three steps and ignored the fact that the dot pattern continued, for example, "The answer

Incorrect—The work is completely incorrect or irrelevant, or the response states, "I don't know."

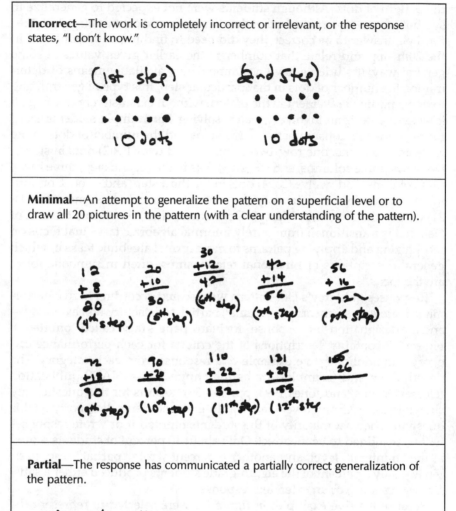

(1st step)
. . . . .
. . . . .
10 dots

2nd step
. . . . .
. . . . .
10 dots

Minimal—An attempt to generalize the pattern on a superficial level or to draw all 20 pictures in the pattern (with a clear understanding of the pattern).

$$
\begin{array}{c} 12 \\ +8 \\ \hline 20 \end{array}
\quad
\begin{array}{c} 20 \\ +10 \\ \hline 30 \end{array}
\quad
\begin{array}{c} 30 \\ +12 \\ \hline 42 \end{array}
\quad
\begin{array}{c} 42 \\ +14 \\ \hline 56 \end{array}
\quad
\begin{array}{c} 56 \\ +16 \\ \hline 72 \end{array}
$$
(4th step) (5th step) (6th step) (7th step) (8th step)

$$
\begin{array}{c} 72 \\ +18 \\ \hline 90 \end{array}
\quad
\begin{array}{c} 90 \\ +20 \\ \hline 110 \end{array}
\quad
\begin{array}{c} 110 \\ +22 \\ \hline 152 \end{array}
\quad
\begin{array}{c} 131 \\ +24 \\ \hline 155 \end{array}
\quad
\begin{array}{c} 155 \\ 26 \end{array}
$$
(9th step) (10th step) (11th step) (12th step)

Partial—The response has communicated a partially correct generalization of the pattern.

When the pattern starts with 2 dots the next step is to add 4 dots to it and the 3rd step is to add 6 dots to it so every time there is a new step you add 2 dots to the last amount you added on to the last step.

→ you would multiply two dots on and on until you reached the 20th step

Satisfactory—The response contains a completely correct generalization of the pattern but does not include—or incorrectly states—the number of dots in the 20th step.

Multiply each step by 1# higher such as

$1 = 1 \times 2$
$2 = 2 \times 3$
$3 = 3 \times 4$
$4 = 4 \times 5$

Extended—This response contains a completely correct generalization of the pattern and specifies that there are 420 dots in the 20th step.

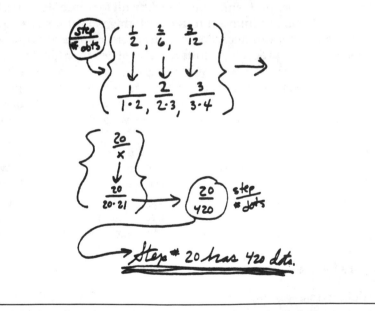

Fig. 9.2 Marcy's Dot Pattern: Performance categories and sample responses
Source: Dossey, Mullis, and Jones. *Can Students Do Mathematical Problem Solving? Results from Constructed-Response Questions in NAEP's 1992 Mathematics Assessment,* 1993

is 20 because I added 2 + 6 + 12 and got 20." As was the case for fourth-grade students on the Extend Pattern problem discussed earlier, it appears that eighth-grade students also have difficulty dealing with the concept of continuity in patterns.

Despite the high percentage of poor responses in the sample gathered for this chapter, there were some responses that illustrated ways in which students were attempting to generalize the pattern of dots. Figure 9.3 contains examples of these responses. Responses 1–4 in the figure represent the variety of ways students explained how to find the number of dots in the 20th figure. Of particular interest is the rather sophisticated use of a "formula" for the pattern of differences used in response 1. Responses 2–4 illustrate how some students gave more detail than others about why multiplying 20 by 21 was the correct procedure. Responses 5 and 6 are typical of those responses in which students used a "brute force" method to determine that there are 420 dots in the 20th figure; that is, although they obviously recognized the pattern, some students were compelled to list the remaining 17 steps.

Because NAEP data from the student questionnaire showed that in 1992 only 28 percent of eighth-grade students were enrolled in prealgebra and only 20 percent in algebra, it was not surprising that Marcy's Dot Pattern, a multistep prealgebra problem, would be difficult for those students. Lack of exposure to algebraic thinking through exploring patterns may also have contributed to low performance. Based on teachers' responses to questions about their classroom practices, results reveal that less than half the eighth-grade students had teachers who reported giving heavy emphasis to algebra and functions concepts. Instead, in 1992 more emphasis was given to topics in numbers and operations, as shown by the fact that about three-fourths of the students in grade 8 had teachers who placed heavy emphasis on numbers and operations.

Marcy's Dot Pattern represents the sort of item that provides middle-grades students with an opportunity to demonstrate the extent to which they are capable of thinking in general terms. Teachers could give this item to students and note the extent to which they generalize and the nature of their arguments. Such information could provide teachers with important insights about students' thinking and their understanding of patterns.

Informal Algebra

In addition to assessing elementary and middle school students' grasp of patterns and functions, NAEP also assessed their understanding of algebra topics presented informally, such as interpreting number sentences in the form of equalities or inequalities, translating from verbal to symbolic form, and graphing points in the coordinate plane. Students develop

Continued

Response 3

Marcie should get the number
of dots by multiplying the step
by one # that is more so 3's.
Answer- 420

Response 2

(1st step)   (2nd step)   (30th step)

$1 \times 3 = 3$ $2 \times 3 = 6$ $\begin{array}{r}30\\ \times\ 3l\\ \hline \end{array}$

(430 dots)

each step adds 1 row and 1 dot to the length
the height of rows ~~will~~ as equivalent to the
step number and the number of dots in a row
is 1 number more than the step number.

Response 1

In each consecutive step, the number of
dots to add goes up by 2.
Thus the formula would be 2 + 4 + 6 + 8...

This is a
formula for   (40+2) 20 ÷ 2
adding portions   = (42)20 ÷ 2   answer: 420
         = 840 ÷ 2
         = 420

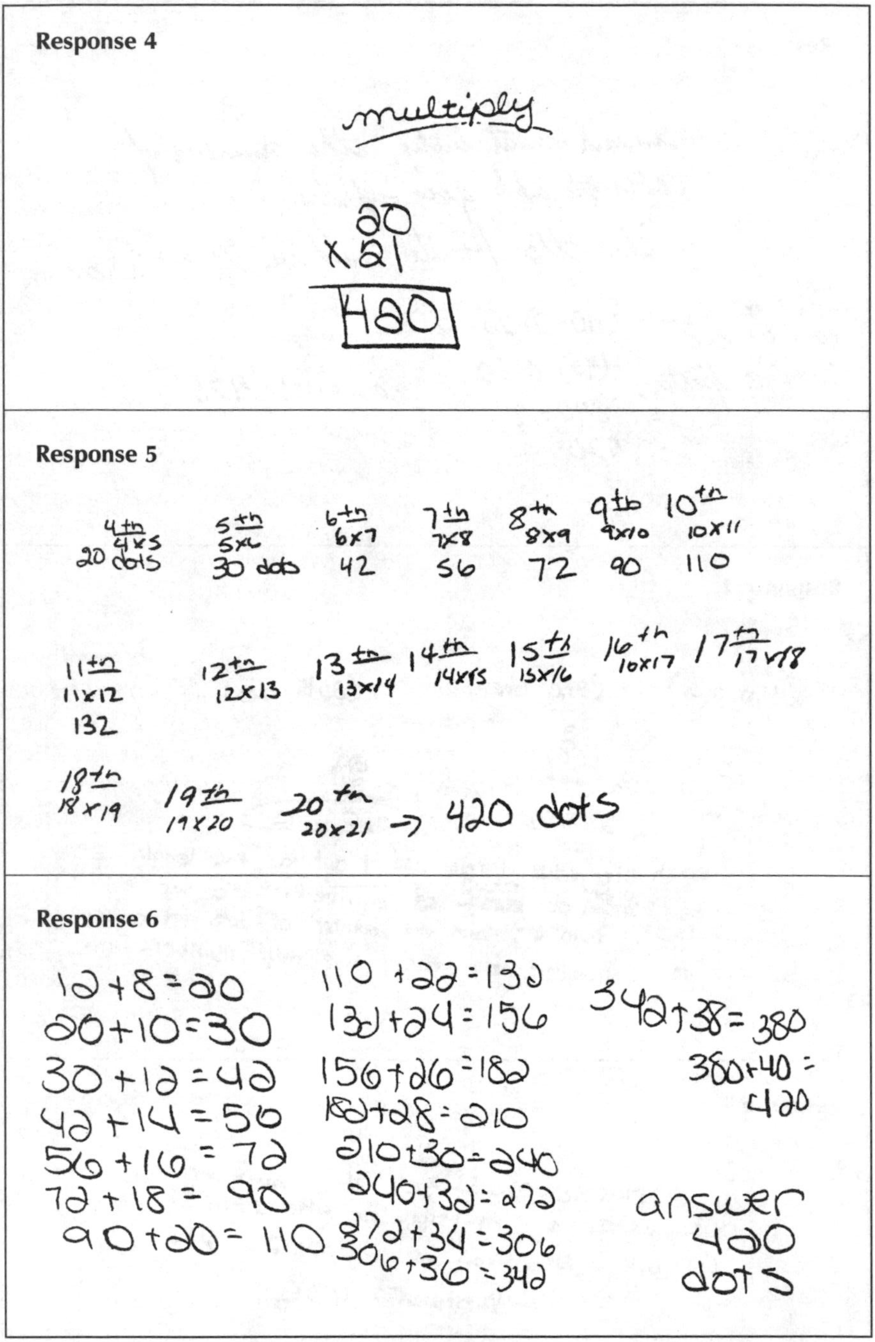

Response 4

multiply

$$30 \times 16$$
$$\boxed{480}$$

Response 5

4th	5th	6th	7th	8th	9th	10th
4×5	5×6	6×7	7×8	8×9	9×10	10×11
20 dots	30 dots	42	56	72	90	110

11th	12th	13th	14th	15th	16th	17th
11×12	12×13	13×14	14×15	15×16	16×17	17×18
132						

18th	19th	20th
18×19	19×20	20×21 → 420 dots

Response 6

10 + 8 = 80
20 + 10 = 30
30 + 12 = 42
40 + 14 = 56
50 + 16 = 72
70 + 18 = 90
90 + 20 = 110

110 + 22 = 132
130 + 24 = 156
156 + 26 = 180
180 + 28 = 210
210 + 30 = 240
240 + 32 = 272
272 + 34 = 306
306 + 36 = 342
342 + 38 = 380
380 + 40 = 420

answer
420
dots

Fig. 9.3 Marcy's Dot Pattern: Additional student responses

these sorts of capabilities in elementary and middle school mathematics as well as in prealgebra courses, so fourth- and eighth-grade students typically have encountered these topics in mathematics classrooms.

Equality and Inequality

Students in the elementary grades often informally experience equations and inequalities in the context of number sentences that contain symbols (for example, a blank or a box) rather than letters. A typical item of this type is, What number should go in the box to make this number sentence true: $3 + \square = 7$? Three NAEP items involved equations or inequalities in number sentence form, and results on these items showed that performance levels improved among grade levels. In particular, percent-correct values for fourth-grade students ranged from 20 percent to 57 percent; for eighth-grade students from 58 percent to 78 percent; and for twelfth-grade students from 78 percent to 83 percent. In general, the number sentences that involved finding several solutions or all possible solutions to inequalities were more difficult than finding a number that satisfied an equation. For example, on a regular constructed-response item that asked eighth-grade students to write two numbers that make the inequality $54 < 3 \times \square$ true, 49 percent supplied two correct numbers and another 11 percent produced only one correct number. On another inequality item in multiple-choice format, 55 percent of the fourth-grade students, 25 percent of the eighth-grade students, and 11 percent of the twelfth-grade students chose the distracter representing a single answer rather than a range of answers, which suggests that those students were treating the inequality as an equation. However, the number of students seemingly ignoring the inequality in that item declined substantially by grade 12.

Representing Verbal Statements in Symbolic Form

Three items assessed translation from verbal statements to symbolic form. Students did well on these items, with percent-correct values ranging from 48 percent to 64 percent in grade 4 and from 57 percent to 62 percent in grade 8. For example, on a released item that asked students to choose the correct representation for the total number of newspapers delivered in 5 days if "$\square$" represents the number of papers delivered each day, about half the fourth-grade students and more than 80 percent of the eighth-grade students chose the correct answer of $5 \times \square$. The most popular incorrect answer among the students in grade 4 was the one representing an additive rather than a multiplicative expression.

Graphing on the Number Line and in the Coordinate Plane

Graphing skills develop concurrently with work on other informal algebra topics. Four items assessed fourth- and eighth-grade students' capabilities with identifying points on the number line and locating points in the coordinate plane. Performance results from a regular constructed-response item show that about half the fourth-grade students could identify coordinates of points on a number line as the first part of a two-step problem. From their success one can infer that most of those students were familiar enough with the number line to find the correct coordinates in the first step of the item. Unfortunately, the structure of the item does not provide information regarding the number of students (in addition to those who were correct) who correctly located points but who did not complete the second step correctly.

Results for students in grade 8 showed that about one-half to three-fourths of the students correctly answered more-complex graphing items, in particular, items that involved identifying or locating points in a coordinate system. One secure, regular constructed-response item administered only to eighth-grade students required them to locate a point in the coordinate plane that satisfied certain conditions. Slightly more than 40 percent of the students located the point and correctly described its coordinates. Performance by students in grade 8 was higher on another item that asked them to choose the coordinates of the remaining vertex of a regular polygon, given the coordinates for the other vertices. Just about half the eighth-grade students and three-fourths of the twelfth-grade students answered that item correctly. The performance difference for students in grade 8 between the two items just described could likely be attributed to the format of the items, with higher performance on the multiple-choice question than the constructed-response question, or to the fact that for the item requiring location of a point in the coordinate plane, students had to translate from a verbal description to a graphical representation.

The item shown in table 9.6 affords the opportunity to examine students' performance on graphing in the coordinate plane between two grade levels. Here, with (4, 4) provided as an example, students had to identify two other points in the plane that had coordinates of the form (x, x). About half the fourth-grade students and more than 80 percent of the eighth-grade students circled either two correct points or one correct point and no others. However, the way in which this item was presented may have maximized the likelihood of correct responses because—owing to the equality of the coordinates—students did not have to distinguish between the first and second coordinates to circle points that fit the given condition.

Looking at a set of student responses to this regular constructed-response item provided additional information about which points were most often circled by students in each grade and about misconceptions those students had about the concept of points with equal first and second coordinates. The fourth- and eighth-grade samples consisted of approximately 800 responses and about 160 responses, respectively. The correct responses in

Table 9.6 Coordinate-Graph Item

Item	Percent Responding	
	Grade 4	Grade 8

On the grid below, the dot at (4, 4) is circled. Circle two other dots where the first number is equal to the second number.

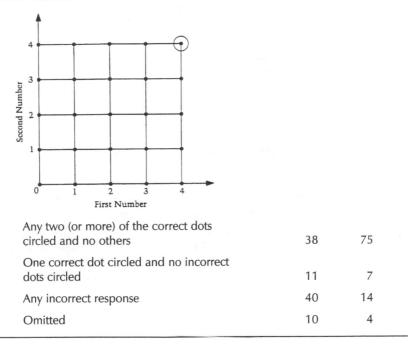

Any two (or more) of the correct dots circled and no others	38	75
One correct dot circled and no incorrect dots circled	11	7
Any incorrect response	40	14
Omitted	10	4

Note: Percents may not add to 100 because of rounding.

which one or two points were circled reveal that students in both grades most often circled the points (3, 3) or (2, 2) or both, which were the points closest on the diagonal to the given point (4, 4). The points with the lowest frequency of selection were (1, 1) and (0, 0), the points farthest away from the given point. This suggests that the sample point provided in the item may have influenced the points students selected. With respect to the types of incorrect points circled most often, the most common error made by fourth-grade students and eighth-grade students was to select points of the form (4, y), with (4, 0) having the highest selection frequency. It is possible that some students chose points of the form (4, y) because the item asked for "two other dots where the first number is equal to the second number" and points of the form (4, y) were ones for which the first number, 4, of this point was equal to the second number, 4, of the given point.

Performance on the NAEP graphing items suggests that a substantial number of students do not demonstrate fairly rudimentary graphing skills involving location of points on the number line and in the coordinate plane. About half of the students in grade 4 could locate points on the number line or in the plane when quite simple conditions determined their locations. In grade 8 fewer than half of the students could locate points in the plane that satisfied a particular condition, and even at grade 12, only three-fourths of the students could do so. When the requirements of problems include taking into account some constraint(s) on a point's location, a substantial number of students are unable to correctly locate or identify coordinates of such points in the plane. Instruction in graphing needs to provide students with more than point-plotting experiences. Students need to determine relationships that exist between and among points and to reason about the effects various constraints have on the location of points on the number line or in the plane.

TRADITIONAL ALGEBRA TOPICS

A substantial portion of the NAEP algebra and functions items dealt with traditional algebra topics such as simplifying and evaluating expressions, solving equations and inequalities, evaluating and graphing functions, and solving and graphing systems of equations. In contrast to the informal algebra items, these relied more on students' capabilities to work symbolically with variables, equations, and inequalities; to use function notation and recognize graphs of functions; and to use symbolic descriptions of patterns when generalizing.

Simplifying, Evaluating, and Writing Expressions

Seven NAEP items assessed students' ability to simplify, evaluate, and write numeric and algebraic expressions, with most items administered to students in grade 12. The majority of those students were successful in evaluating simple numeric and algebraic expressions, with percent-correct values ranging from about 50 percent to 95 percent. However, performance levels dropped to 40 percent correct and lower for items that involved somewhat complex order of operations, evaluation of terms in a sequence, and cubic expressions. Table 9.7 contains examples of released items involving algebraic expressions.

Results from a secure pair of order of operations items show that over 90 percent of students in grades 8 and 12 could evaluate a simple numeric expression requiring one use of the order of operations rules. However, when asked to simplify the expression in item 1 in table 9.7 and choose the correct answer, only 22 percent of the students in grade 8 and 29 percent in grade 12 produced the correct value. Because of that item's multiple-choice

Table 9.7 Simplifying and Evaluating Algebraic Expressions

	Percent Responding	
Item	Grade 8	Grade 12
1[a]. $3^3 + 4(8 - 5) \div 6 =$		
A. 6.5	26	40
B. 11	10	10
C. 27.5	15	5
D. 29*	22	29
E. 34.16	22	15
2. $\dfrac{6 \times 10^3}{3 \times 10^5} =$		
A. 0.5×10^2	—	6
B. 2×10^2	—	27
C. $2 \times 10^{0.6}$	—	9
D. 0.5×10^{-2}	—	6
E. 2×10^{-2} *	—	48
3. If $x = -4$, the value of $-4x$ is		
A. −16	—	9
B. −8	—	7
C. 8	—	7
D. 16*	—	75

*Indicates correct response.
[a]Item is from an item block for which students were provided with, and permitted to use, scientific calculators.
Note: Percents may not add to 100 because of rounding or omissions.

format, it is impossible to know with certainty how many students merely guessed and how many used incorrect order of operations rules. In both grades, the most frequently chosen distracter was choice A (6.5), an answer likely produced either by performing the operations from left to right as presented in the problem (adding 27 and 12 and then dividing by 6), or by getting 9 when cubing 3, then adding 4, multiplying by 3, and dividing by 6. The increase in this error from grade 8 (26 percent) to grade 12 (40 percent) is particularly distressing and difficult to explain. One possible explanation involves students' lack of experience with such items in high school mathematics courses, including lack of attention to how a calculator can be used to evaluate such numeric expressions. While working on this item, students were provided with, and permitted to use, a scientific calculator. However, even with a calculator available, many students were unable to evaluate this moderately complicated expression requiring several decisions about order of operations.

As illustrated previously, incorrectly evaluating exponential expressions is a common error. Item 2 in table 9.7 required students to simplify a quotient given in scientific notation, and about half the twelfth-grade students selected the correct answer. However, 27 percent chose the distracter 2×10^2, perhaps from dividing 6 by 3 and 10^5 by 10^3 or from applying an incorrect generalization such as

$$\frac{10^m}{10^n} = 10^{|m-n|}.$$

Another kind of exponential error pattern appeared in a secure NAEP item that required twelfth-grade students to simplify an exponential expression of the form $(cx^m)^n$. Although 44 percent of the students responded correctly, another 45 percent chose responses that either corresponded to multiplying the coefficient by the power (that is, cn rather than c^n) or to adding the exponents (that is, $x^{(m+n)}$ instead of x^{mn}) or to both kinds of exponent errors. The three preceding items highlight the difficulties students have with simplification and evaluation of expressions, particularly those involving exponents, despite the fact that many algebra courses place heavy emphasis on such skills.

Two items administered to students in grade 12 involved substitution of values into symbolic expressions. Most of those students had little difficulty correctly answering one of these items—item 3 in table 9.7. Seventy-five percent chose the correct answer, 16, with fewer than 10 percent selecting the most obvious distracter, –16. Another item, this one in regular constructed-response format, required twelfth-grade students to use whole numbers to find the first few terms of a sequence described by a particular algebraic expression, and this item was somewhat more difficult. Only 40 percent of the twelfth-grade students correctly and completely answered the question by identifying all appropriate numbers. Another regular constructed-response item asked students to write expressions, in terms of a given variable, that represented two quantities in a word problem. This was quite difficult for most students—only 9 percent wrote two correct expressions, and an additional 2 percent correctly represented only one of the quantities. Although students in grade 12 appear to be successful in simplifying and substituting values into simple expressions, more-complicated expressions create difficulties, and writing symbolic expressions from verbal statements is even more difficult.

Writing, Solving, and Graphing Equations and Inequalities

Eighteen NAEP items administered to students in grades 8 and 12 involved equations, systems of equations, and inequalities. Nine items addressed writing and solving equations, five items involved systems of equations, and four items involved solving and graphing inequalities. Table 9.8 summarizes eighth- and twelfth-grade students' performance on those items. On the set of items involving equations and inequalities, percent-correct values for eighth-grade students ranged from about 30 percent to about 45 percent. On a much larger set of items administered only to students in grade 12, including items on systems of equations, twelfth-grade students' performance varied widely, ranging from highs of about 60 to 85 percent on simple items to lows of about 10 to 35 percent on items presented in regular constructed-response format (e.g., items 8 and 13) and items that assessed more-complicated equations and inequalities (e.g., items 17 and 18). The three sections that follow separately discuss the items and results for the topics of equations, systems of equations, and inequalities.

Writing and Solving Equations

On the nine items that addressed writing and solving equations, only when the equations were very simple did the majority of students find the correct solution. When the equations had more-complicated coefficients or were nonlinear equations, performance dropped considerably. For example, items 3 and 9, described in table 9.8, assessed performance on determining or writing an equation from a table of values or a verbal description. For item 3, only about one-third of the eighth-grade students were able to choose the correct linear equation when given a table of values, although almost 60 percent of the twelfth-grade students were successful on this item. Writing an equation that reflected the information and relationship in a word problem was much more difficult, as shown by performance results for item 9, a regular constructed-response item—only 14 percent of twelfth-grade students wrote a correct two-variable linear equation that described what was given in a word problem. Producing an equation that models information in a problem is considerably more difficult for students than choosing a solution from ones that are given to them.

A closer look at results for two released items, shown in table 9.9 may shed some light on students' performance in solving equations. Item 1 was a regular constructed-response item that required twelfth-grade students to solve a simple quadratic equation, and item 2 was a multiple-choice item that required selecting from a list of choices the value of x that satisfies an exponential equation. On both items students were provided with, and permitted to use, scientific calculators. Two-thirds of the twelfth-grade students correctly solved the quadratic equation, but only about one-third

Table 9.8 Performance on Items Involving Equations and Inequalities

| | Percent Correct | |
Item Description	Grade 8	Grade 12
Equations		
1. Solve a linear equation in two variables for one variable in terms of the other.	44	—
2. Determine which ordered pair is a solution to a linear equation in two variables.	39	—
3. Determine which equation in two variables fits the values given in a table.	33	59
4. Solve a simple quadratic equation.	—	67
5. Determine which set of ordered pairs is a solution to a nonlinear equation.	—	48
6. Choose the correct solution of a literal equation in three variables for a specified variable.	—	37
7. Determine the solution of an exponential equation.	—	34
8. Solve a linear equation with exponential and radical coefficients.	—	26
9. Write an equation for a word problem.	—	14
Systems of Equations		
10. Select the solution to a pair of simple linear equations.	—	85
11. Use substitution and select the values for several simple quadratic equations.	—	52
12. Solve a system of equations in two variables, and select the value for one of them.	—	48
13. Solve a system of two linear equations in two variables.	—	26
14. Graph a system of two linear equations in two variables.	—	10
Inequalities		
15. Graph a simple inequality on a number line.	38	67
16. Choose the correct graph for a linear inequality.	31	60
17. Choose the correct solution for a quadratic inequality.	—	33
18. Choose the graph corresponding to a linear absolute value inequality.	—	27

Note: Items 4, 8, 9, 13, 14, and 15 were regular constructed-response items, and the rest were multiple-choice items.

Table 9.9 Solving Equations

Item[a]	Percent Responding Grade 12
1. If $n \times n = 729$, what does n equal?	
Correct answer of 27	67
Any incorrect answer	26
Omitted	7
2. For what value of x is $8^{12} = 16^x$?	
A. 3	6
B. 4	26
C. 8	16
D. 9*	34
E. 12	10

[a]Items are from item blocks for which students were provided with, and permitted to use, scientific calculators.

*Indicates correct response.

Note: Percents may not add to 100 because of rounding or omissions.

selected the correct answer for the exponential equation. In each case, however, NAEP results provide no information on exactly how students solved the equation. Student responses were not examined for item 1; however, it is likely that some students recognized that $n \times n = n^2$ and used the calculator to find the square root of n. On that item no information is available concerning the extent to which the errors made by 26 percent of the students included squaring 729. In the instance of the multiple-choice item, in addition to guessing, students may have resorted to substituting the values given in the choices into the exponential equation instead of applying an understanding of exponential equations by transforming the equation so that exponents of like bases could be equated.

As reported previously, both items in table 9.9 were in calculator blocks. The percent-correct value for students who reported using the calculator on item 1 was 78 percent, in contrast to only 30 percent correct for those who reported not using the calculator, a difference significant at the .05 level. If students realized that taking the square root of 729 would yield the solution, using the calculator could make this a relatively easy calculation. On item 2, 67 percent of those reporting calculator use were correct, whereas only 14 percent of those reporting no calculator use were correct. This difference also was significant at the .05 level. Although calculator use might have been responsible for improved performance, one must keep in mind that another explanation for the performance differences on both of these items could be that the more capable students were those more likely to choose to use the calculator. It is not clear from NAEP results how students used the calculator in solving the exponential equation, although one possibility is that they used the calculator and the answer choices A through E to compute 16^3, 16^4, 16^8, 16^9, and 16^{12}, comparing values derived from each answer choice to their previously calculated value of 8^{12}. Another possible use of the calculator would be to use a guess-and-test strategy, selecting a trial value for x, computing 16^x, and then systematically adjusting values for x until 16^x equals 8^{12}. With this item it might be interesting to omit the choices and observe students' work, asking them to describe their solutions, including ways in which they may have used the calculator.

Students had difficulty solving equations when the item required them to solve for one variable in terms of another or when coefficients were nonintegral expressions. On a secure item involving the manipulation of a simple linear equation, 44 percent of eighth-grade students chose the correct solution for one variable in terms of the other. Although only 37 percent of twelfth-grade students chose the correct solution for another item involving expression of a specified variable in terms of others, the equation in that item was a more complicated literal equation that involved three variables. Solving a linear equation with more-complicated coefficients was equally difficult; just over one-fourth of the twelfth-grade students correctly solved a regular constructed-response item that asked them to find the value of N, rounded to the nearest tenth in the equation $\sqrt{8} N = 3^5$. Despite the fact that this was a linear equation of the simple form $ax = b$, it was complicated by a radical coefficient and constant term that entailed evaluation of an exponential expression. Consequently, only 26 percent of the twelfth-grade students solved it correctly and 10 percent omitted this item.

In addition to items that involved students' solving the equations just described, two secure items assessed students' abilities to determine whether ordered pairs were solutions to two-variable equations. Nearly 40 percent of eighth-grade students selected from given ordered pairs one that was a solution to a two-variable linear equation. Twelfth-grade students were slightly more successful, using substitution as the presumed method of solution, on a different item that required them to select a set of ordered pairs that satisfied a simple nonlinear equation. Nearly half of those students chose the correct set, with an additional one-fourth incorrectly choosing the set of pairs reflecting the same relationship but with the x- and y-values reversed. Students' responses to these items suggest that close to half of the students in grades 8 and 12 can choose solutions from a list of potential solutions to relatively simple equations, but because of the multiple-choice format of those items, these results may have been influenced by guessing.

Important information concerning students' understanding of ordered pairs and the nature of solutions of equations would be available if additional NAEP items in regular constructed-response format had addressed these topics. In addition to knowing students' performance on solving equations or choosing from among potential solutions, it is important for teachers and researchers to learn more about what students think constitutes a solution and their understanding of what constrains the values one obtains when solving a particular equation.

Systems of Linear Equations

Five NAEP items assessed twelfth-grade students' ability to solve systems of linear equations. Performance levels varied, with percent-correct values ranging from 85 percent to a low of 10 percent on items that required students to graph a system of two linear equations in two variables. When given a system of two simple linear equations in one variable, 85 percent of twelfth-grade students determined the value for the variable. However, because the item format was multiple choice, it was possible for students to substitute the values given in the choices into the equations to determine the solution. On another item, more than half of the students in grade 12 also were successful in substituting values into a system of simple quadratic equations.

Two NAEP items involved solving systems of linear equations in two variables. Students could not solve these equations directly by substituting the given responses into the system because one item was in regular constructed-response format and the multiple-choice item gave potential solutions for only one of the variables. Because nearly half of the twelfth-grade students correctly solved the multiple-choice item and about another one-fourth of the students chose the distracter representing the correct solution for the other variable, one could conclude that nearly three-fourths of the students in grade 12 may be able to solve a two-variable linear system. However, on the regular constructed-response item 26 percent of the students correctly solved for both variables. Even if one includes the 6 percent who gave the correct solution for only one of the two variables in the system, only 32 percent of the twelfth-grade students correctly solved for at least one of the variables in the system. In addition, many of the 21 percent who omitted this item were probably unable to solve the system. Also, because this item was nearly the last item in its item block, 36 percent of those taking the test did not reach this item. If one assumes that only the more capable students reached this item, one might expect that in the total population somewhat fewer than 32 percent of twelfth-grade students could solve this system correctly. The results from these two items in different formats differ greatly; consequently, it is not clear whether most students in grade 12 can solve simple linear systems.

Although some twelfth-grade students were able to symbolically solve a two-variable linear system, it appears that a much smaller percentage can solve such a system by graphing. In a regular constructed-response item, students in grade 12 were asked to graph a system of linear equations and give the coordinates of the solution to the system. Only 10 percent of the students produced a completely correct solution, another 11 percent provided the correct solution without completing the graph, and 16 percent omitted the item. Students' performance on this item combined with the percent-correct values on the items just discussed suggests that whereas more students may be able to produce a correct solution symbolically, perhaps only a small percentage of them use graphical approaches. Students' difficulty with the graphical solution item may be a result of curricular emphasis being placed on symbolic solutions to systems (e.g., substitution) at the expense of graphical solutions. Graphical approaches often are used to introduce solutions of systems of equations, but emphasis soon shifts to finding solutions symbolically. Students, particularly when graphing calculators are available, need to be encouraged to continue to use graphical solutions and to translate between and among graphic, numeric, and symbolic representations. It will be interesting to note whether, as graphing calculators are used more widely, students become more proficient with graphical solutions to systems, particularly if they have had regular access to graphing tools.

Solving and Graphing Inequalities

Three NAEP items focused on linear, quadratic, and absolute-value inequalities. One item required eighth- and twelfth-grade students to select the correct graph of the solution of a linear inequality of the form $ax + b \geq c$. Although only 31 percent of students in grade 8 answered correctly, performance improved to 60 percent correct in grade 12.

Two items, one involving a quadratic inequality and the other a linear absolute-value inequality, were much more difficult for twelfth-grade students. Only 33 percent of students in grade 12 correctly identified the solution to a quadratic inequality, and only 27 percent selected the correct graph of a linear absolute-value inequality. Quadratic inequalities are often difficult for students, but because the item involving the quadratic inequality was in multiple-choice format, the choices had the potential to provide clues to the solution. For example, because responses were given either in the form $a \leq x \leq b$ or in the form $x \leq a$ or $x \geq b$, students who understood the meaning of a solution set could substitute values from each of the choices into the inequality and eventually arrive at a correct solution. Yet, even with that capability available, only one-third of the students successfully answered that item. Twelfth-grade students' rate of success on choosing the correct graph for the absolute-value inequality was only slightly better than the level of guessing for a five-choice item (27 percent correct), and the popularity of the distracter associated with disregarding the absolute value (30 percent) suggests that many students may have solved the inequality by simply ignoring the absolute-value symbol.

Functions

In addition to items that addressed functions informally in the context of patterns and relationships, eight NAEP items explicitly addressed more-formal aspects of functions. These items assessed students' ability to evaluate functions, interpret the graphs of functions, identify graphs of particular functions, and find zeros of a function. Although performance on those eight items ranged from about 20 percent correct to 75 percent correct, seven of the eight items had percent-correct values below 60 percent. Students were most successful on items that involved evaluating functions or interpreting information from the graphs of functions, and they were least successful when asked to identify particular functions.

Evaluating Functions

Two items, shown in table 9.10 and administered to students in grade 12, addressed finding function values when given a function rule in symbolic form. Item 1 asked twelfth-grade students to evaluate a quadratic function, and item 2 asked them to evaluate the composition of a linear and a quadratic function. In the first item, nearly 40 percent of the students computed the correct value for the quadratic function given a decimal input value, but nearly one-fourth omitted the item. While the reasons why those students omitted the item cannot be identified with certainty, one possible reason is that they were unfamiliar with function notation. Surprisingly, however, nearly 60 percent of the students in grade 12 chose the correct value for the composition of two functions in item 2 and almost no students omitted this seemingly more difficult item. If students were unfamiliar with function notation, it seems unlikely that they would perform so well on item 2, so other explanations for the performance difference between items 1 and 2 are warranted.

One explanation may be that item 1 was a constructed-response item, often a more difficult format than multiple choice. Another explanation may be that decimal values were required in item 1, whereas integer

Table 9.10 Symbolic Representations of Functions

	Percent Responding
Item	Grade 12
1[a]. If $f(x) = 4x^2 - 7x + 5.7$, what is the value of $f(3.5)$?	
Correct response of 30.2	39
Any incorrect response	39
Omitted	21
2. If $f(x) = \dfrac{2x + 1}{3}$, and $g(x) = 2x^2 + 2$, then $f(g(2)) =$	
A. 3	5
B. 5	13
C. 7*	59
D. $7\frac{5}{9}$	15
E. $16\frac{2}{3}$	9

*Indicates correct response.

[a]Item is from an item block for which students were provided with, and permitted to use, scientific calculators.

Note: Percents may not add to 100 because of rounding or omissions.

values resulted in item 2. Also, because item 2 was positioned near the end of the item block, 40 percent of the students did not reach it and left their paper blank. On the basis of this rather high "not reached" rate, we may conclude that only better-performing students reached this item and answered correctly. It is possible that students who did not understand function notation may have guessed or may have chosen "7" simply because it was one of the three integer choices.

Item 1 was a calculator-active item in a calculator block, and the self-report calculator use results are somewhat striking. Nearly three times as many students reported using a calculator as reported not using one, and of those twelfth-grade students who reported that they used a calculator, 61 percent were correct, versus only 6 percent correct for those who reported not using the calculator. This difference was significant at the .05 level. Although it is not possible to conclude that higher performance can be attributed to calculator use, simply because more capable students may have been the calculator users, it is at least possible that use of the calculator for the decimal computations made the calculations easier and more likely to be correct.

In spite of inconsistent performance on items that asked students to evaluate functions from symbolic rules, a reasonable number of students were able to evaluate a nonlinear function and a composition of linear and nonlinear functions. This is particularly noteworthy given that students who had not taken advanced mathematics courses may have had little or no exposure to function notation.

Interpreting and Identifying Graphs of Functions

Three items addressed interpreting graphs of functions. Performance on the these items was mixed, ranging from 75 percent correct to 25 percent correct, with the majority of items below 50 percent correct. One item, set in the context of a race involving two runners, combined concepts from data analysis and functions and required students to interpret information depicted in two superimposed graphs of functions. About two-thirds of the eighth-grade students and three-fourths of the twelfth-grade students correctly determined the time at which one runner passed the other. One possible explanation for students' success on this item is that some students who may not have interpreted the graphs correctly may have viewed the graphs as pictures (Clement 1985) giving the paths of the two runners. Consequently, students could have interpreted the graphs' intersection point—namely, where the "paths" crossed—as the point at which one runner passed the other. In doing so, students would have obtained the correct answer from an *incorrect* interpretation of the graph. Also, some students who simply may have been looking for some point to choose for their answer may have correctly selected the intersection because it was different from the other points on the graphs. These students would have obtained the correct answer without *any* interpretation of the graphs of the functions.

In another item, twelfth-grade students were given a graph of a step function that modeled a realistic situation and asked to use two values from the graph to solve a two-step problem. Fifty-two percent solved the problem correctly, 19 percent appeared to have reasoned incorrectly about the information in the problem rather than about the graph itself, and 18 percent appeared to have incorrectly produced a value from the graph. The use of a step function—a less commonly encountered function in the secondary school curriculum than continuous, polynomial functions—combined with the fact that this was a multistep word problem, a complicating factor in many of the NAEP items (see chapter 5), most likely contributed to the item's difficulty.

A more demanding graphical interpretation item was a secure, extended constructed-response question that again involved important concepts in both data analysis and functions. In that item, twelfth-grade students were asked to produce a detailed written explanation about events depicted in a graph. As was the case for most other extended-response questions, performance levels were quite low, with only 9 percent of the students producing extended responses that completely and correctly described each portion of each time interval correctly and another 19 percent producing satisfactory responses that described at least some aspects of the graph correctly. The remaining three-fourths of students, however, wrote responses that were only partially or minimally correct or that were incorrect. These results suggest that doing *and* communicating about mathematics is more difficult for students than merely doing mathematics.

Another item, shown in table 9.11, asked students to identify the graph of a particular type of function—the composition of the absolute-value function with another function. Only 20 percent of twelfth-grade students chose the correct graph for $y = |f(x)|$ when given the graph of $y = f(x)$, with nearly twice as many students incorrectly choosing the graph depicted in choice A, representing $-f(x)$ rather than $|f(x)|$. This suggests that some students were interpreting absolute value as "opposite."

One secure item presented a slightly different aspect of interpretation of graphs of functions. It presented a graph of a function to twelfth-grade students and asked them a question about the number of zeros of the function. Forty-five percent responded correctly, but 34 percent chose distracters that suggested they confused zeros with the vertical axis intercept.

In general, students could interpret graphs more easily than they could identify particular functions or whether a graph represented a function. It is likely that students could profit from additional work with graphical representations of functions. The 1992 NAEP provides no evidence of students' capabilities to evaluate or graph functions with tools such as graphing calculators, so it is not clear how capable students might be if they had access to a graphing calculator for the NAEP items.

ALGEBRAIC REASONING

As shown by NAEP results, algebraic reasoning or algebraic thinking (Wagner and Kieran 1989)—reasoning about the nature of objects (as described by Sfard [1991]) such as expressions, equations, functions, and graphs and conjecturing and drawing conclusions about relationships between or among them—was difficult, even for older students. Results from four items, two secure and two released, provide insight into twelfth-grade students' capabilities with making and testing conjectures and using symbolic representation to describe general results, key components of alge-

Table 9.11 Function Involving Absolute Value

Item	Percent Responding Grade 12

The figure above shows the graph of $y = f(x)$. Which of the following could be the graph of $y = |f(x)|$?

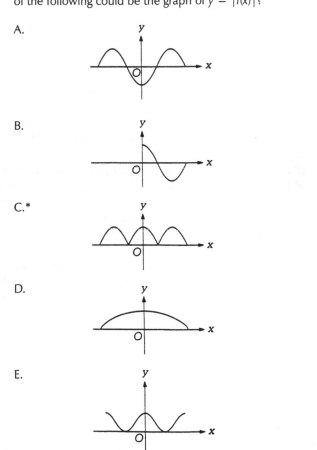

A.		36
B.		18
C.*		20
D.		7
E.		16

*Indicates correct response.

Note: Percents may not add to 100 because of rounding or omissions.

braic reasoning. Two items were presented in regular constructed-response format, and the other two were extended constructed-response questions. Because of the types of data they generate, these open-ended formats have greater potential than multiple-choice items to provide rich information about students' algebraic thinking and reasoning.

Performance on the regular constructed-response items was somewhat uneven. On one of the items, more than half the twelfth-grade students found an appropriate numerical counterexample when asked to disprove a conjecture about numbers in a sequence. Because no actual student responses were examined for this chapter, it is impossible to know with certainty how those students arrived at their responses. It is reasonable to think that students may have been able to get a correct answer by producing the initial terms in the sequence and checking each of them to determine whether they exhibited the given characteristic rather than reasoning in general about the nature of the terms in the sequence. For the other item, which combined number theory concepts and algebraic expressions, when given integer-valued variables and asked to reason about the nature of an expression using those variables under different constraints, slightly more than one-third of students in grade 12 were able to draw correct conclusions about all of the expressions. Another 53 percent of them were able to arrive at conclusions for some expressions, although students had a 50 percent chance of guessing correctly for each expression.

Two released extended constructed-response questions required students either to construct an algebraic proof to explain why a given statement was always true or to determine whether it was possible for a statement to be true. The first of these two extended questions, hereafter referred to as Patterns of Squares and shown in table 9.12, required twelfth-grade students to construct a proof based on elementary algebra and number theory to explain why a statement about the squares of positive integers that end in 5 is always true. Students were provided a scientific calculator to use while answering this question. To answer correctly, students had to use their knowledge of algebra to expand correctly the expression given in the hint ($[10n + 5]^2 = 100n^2 + 100n + 25$). Working with that expression, students then would have to reason that because $100n^2$ and $100n$ are both multiples of 100, their sum is also a multiple of 100 (that is, its tens and units digits are both 0) and then concluded that adding 25 to a multiple of 100 would always result in that number's tens and units digits being 2 and 5, respectively.

Results for this question are also shown in table 9.12, with descriptions of each performance category and a sample response of each category appearing in figure 9.4. The descriptions and sample responses are as they appeared in a NAEP publication (Dossey, Mullis, and Jones 1993, pp. 152–55. Twelfth-grade students had great difficulty with this question. Seventeen

Table 9.12 Patterns of Squares

	Percent Responding
Question[a]	Grade 12

[General directions]

This question requires you to show your work and explain your reasoning. You may use drawings, words, and numbers in your explanation. Your answer must be clear enough so that another person could read it and understand your thinking. It is important that you show all your work.

$$15^2 = 225$$
$$25^2 = 625$$
$$35^2 = 1225$$

The examples above suggest the following statement.

When a positive integer that ends in the digit 5 is squared, the resulting integer ends in 25.

Explain why this statement is always true.

(Hint: $(10n + 5)^2 = ?$)

Extended response	1
Satisfactory response	1
Partial response	1
Minimal response	16
Incorrect	64
Omitted	17

[a]Question is from an item block for which students were provided with, and permitted to use, scientific calculators.

percent of the students omitted the question, and nearly two-thirds either gave responses that were completely incorrect or irrelevant or that stated, "I don't know." Sixteen percent of the students produced minimal responses that contained, with no explanation, the expected algebraic expansion error $(10n + 5)^2 = 100n^2 + 25$. If a partial explanation accompanied the incorrect expansion, then the response was judged to be partial; however, only about 1 percent of the twelfth-grade students' responses were so scored.

Incorrect—The work is completely incorrect or irrelevant, or the response states, "I don't know."

$$(10N+5)^2 = (15_n)^2 = 225$$

If the answer from the equation ends in 5, then the answer will alway be 25.

Minimal—Student provides additional numerical examples only or states $(10n + 5)^2 = 100n^2 + 25$ only.

Any Positive integer times itself will always have 25 at the end of the answer because if, $15^2 = 225$ $15 \times 15 = 225$ it ends in 5 so it will have 25 in the answer

Example $45 \times 45 = 2025$
$5 \times 5 = 25$
thats How you get 25 at the end of each answer

Partial—Student states $(10n + 5)^2 = 100n^2 + 25$, and provides a partially correct explanation.

Because the square of 5 is 25 and the square of 10n is always equal to n squared times 100

ie:
$$(10n+5)^2 = x$$
if $n = 4 \rightarrow [10(n)]^2 = [40]^2 = 1600$
$n^2 = 4^2 = \boxed{16} \times 100 = 1600$
$5^2 = 25$
$1600 + 25 = 1625$

Satisfactory—Student states that $(10n + 5)^2 = 100n^2 + 100n + 25$ and mentions zero(s). The explanation ties 25 to a multiple of 10 or 100.

$$(10n + 5)^2 = 100n^2 + 100n + 25$$

100 times any number or n
leves two empty spases wtch
only the 25 can Fil/

Extended—Student displays a solution that is mathematically accurate and provides a clear and complete explanation.

$$(10n + 5)^2 = 100n^2 + 100n + 25$$

For any Number N,
100N² + 100N, will End
in — 00, which you must

add 25 to. This results
in the Number ending in
25.

Fig. 9.4 Patterns of Squares: Performance categories and sample responses
Source: Dossey, Mullis, and Jones (1993)

Only 2 percent of the students produced either a satisfactory response or an extended response that was mathematically accurate and completely correct. Course-taking patterns among students in grade 12 appeared to have some effect on performance on this question. In particular, of the students who resported studying calculus, 11 percent scored at the satisfactory or extended levels.

Results linking performance on the Patterns of Squares question and answers to the self-report question on calculator use ("Did you use the calculator on this question?") showed that 11 percent of the students who said they used the calculator scored at the satisfactory or extended levels as opposed to 7 percent of those who said they did not use the calculator, with the difference between the percents significant at the .05 level. However, these results provide no information about how students used the calculator as an aid to solving this problem. It is reasonable to think they used the calculator to investigate further the conjecture about the pattern of squares using larger numbers, or they could have used it in more simple ways such as squaring 10 and getting 100.

To attempt to gain additional information about performance on the Patterns of Squares question beyond that available from NAEP publications, a sample of responses representing about 24 percent of the total number of nonblank responses was obtained and analyzed for this chapter. Unfortunately, because of the dearth of mathematically correct responses in the sample, that analysis produced no information that was not already reported in the Dossey, Mullis, and Jones report. The majority of the responses gathered for this chapter were similar to the examples in figure 9.4 for the incorrect and minimal performance categories; that is, the response was completely incorrect or irrelevant, or it consisted only of a few additional examples of squares of numbers ending in 5 (45, 55, 65, and others) or involved an incorrect factorization of the expression given in the hint.

The other extended constructed-response question given to students in grade 12, hereafter referred to as Effective Tax Rates and shown in table 9.13, involved interpreting correctly a definition for an effective tax rate, determining whether two particular effective tax rates are possible, and giving a justification for the decisions. Because the first part of this problem involved setting up and solving an equation, $0.06(x - 10,000) = 0.05x$, this part was deemed as "accessible to most high-school seniors" (Dossey, Mullis, and Jones 1993, p. 140). However, the second part of the problem required an understanding of when an equation has no real solution, that is, $0.06(x - 10,000) = 0.06x$.

Results for this question are also shown in table 9.13, with performance-category descriptions and sample responses appearing in figure 9.5. These descriptions and examples are as they appeared in the Dossey, Mullis, and Jones report (1993, pp. 143–47). Performance levels on the Effective Tax Rates problem were at about the same levels as those for the Patterns of Squares problem just discussed. About one-fifth of the twelfth-grade students omitted the question, and two-thirds gave responses that were judged to be incorrect. Nine percent of the students produced minimal responses that showed some evidence of working with the percents and the monetary amount given in the problem, and another 2 percent produced partial responses that contained a worked-out example of an effective tax rate or a relevant equation. Only 3 percent of the students' responses were judged to be satisfactory or extended.

Again, as was the case for the Patterns of Squares question, calculator use and course-taking patterns seemed to affect performance on Effective Tax Rates, with performance differences even more dramatic. Of students who reported using the calculator, 20 percent scored at the partial level or higher, with only 3 percent of students who reported not using the calculator scoring at or above the partial level. For course taking, 7 percent of students who reported taking precalculus scored at the satisfactory or extended levels, and 17 percent of students who reported taking calculus scored at one of those levels. It is somewhat surprising, however, that students enrolled in higher mathematics did not do better on this problem. Even for students who reported enrollment in calculus, 50 percent pro-

Table 9.13 Effective Tax Rates

	Percent Responding
Question[a]	Grade 12

[General directions]

> This question requires you to show your work and explain
> your reasoning. You may use drawings, words, and numbers
> in your explanation. Your answer must be clear enough so
> that another person could read it and understand your
> thinking. It is important that you show all your work.

One plan for a state income tax requires those persons
with income of $10,000 or less to pay no tax and those
persons with income greater than $10,000 to pay a tax of
6 percent only on the part of their income that exceeds
$10,000.

A person's effective tax rate is defined as the percent of
total income that is paid in tax.

Based on this definition, could any person's effective tax
rate be 5 percent? Could it be 6 percent? Explain your
answer. Include examples if necessary to justify your
conclusions.

Extended response	1
Satisfactory response	2
Partial response	2
Minimal response	9
Incorrect	66
Omitted	20

[a]Question is from an item block for which students were provided with, and permitted to use, scientific calculators.

duced responses that were scored as incorrect, and for students in precalculus, more than 60 percent of their responses were so scored.

Again for the purposes of this chapter, a sample of responses representing about 11 percent of the total number of nonblank responses was obtained and analyzed to supplement information available from NAEP data and publications. Like the Patterns of Squares problem, the set of additional responses to Effective Tax Rates contained very few high-quality responses. However, the analysis uncovered at least one interesting error. To

Incorrect—The work is completely incorrect or irrelevant, or the response states, "I don't know."

1 plan

income $10,000 or less => no tax

income > $10,000 => pay 6% on difference from $10,000

Any person's effective tax rate cannot be 6-percent or be 5 percent.

If everyone does not pay tax because of their income then any person's effective income could not be.

Minimal—Student shows some evidence of working with the 5% or 6% and the $10,000 appropriately.

- Yes, it could be 5% if they made enough money to pay 5% of their total salery.

- Yes, it could be 6%, for the same reason as above.

If you made more than $10,000, you pay tax on what is over 10,000, for example, if you made 20,000, you would pay $600 on the extra $10,000 you made.

Partial—There is evidence of some correct work; i.e., an example of a specific effective tax rate or a relevant equation is displayed.

$10,000 — tax free

- ($100,000 × .05 = $5,000) => Effective Tax

100,000 − 10,000 = 90,000 × .06 = 5400

5,400 is what % of 100,000

$\frac{5400}{100000} = 5.4\%$

If your salary is 100,000 it is possible to be taxed 5% of your total income, even if the 1st 10,000 is tax free.

5% tax
Income = 200,000 × (.05) = 10,000 => Effective tax rate

190,000 × (.06) = 11,400 => taxes paid

6% is possible

Satisfactory—Student correctly shows that the effective tax rate can be 5% OR shows that an effective tax rate of 6% is not possible—but not both.

Let's say Income =
$60,000
−10,000
50,000

$\text{Tax} = .06 (50,000) = 3,000$

$60,000 X = 3,000$

$X = .05 = \boxed{5\%}$ It could be 5%

I don't see how it could be 6%

Extended—The work for both the 5% and 6% effective tax rate cases is clearly and accurately shown.

5% $\leq$ $10,000 no tax

YES $>$ $10,000 6% on over $10,000

$X = $ income over $10,000$

$.06(X) = .05(10,000 + X)$
$.06X = 500 + .05X$
$.01X = 500$
$X = 50,000$

Someone with an income of $60,000 would have effective tax rate of 5%.

NO 6%

$.06(X) = .06(10,000 + X)$
$.06X = 600 + .06X$
$0 \neq 600$

Not possible to have 6% effective tax rate

Fig. 9.5 Effective Tax Rates: Performance categories and sample responses
Source: Dossey, Mullis, and Jones (1993)

solve the problem correctly, understanding the definition of an "effective tax rate" was critical. Yet, nearly 45 percent of the responses in the sample gathered for this chapter showed that students did not have a clear understanding of the given definition. Instead, students focused on the tax rate of 6 percent on the part of income over $10,000 and assumed that this *was* the effective tax rate. A typical response showing this error follows: "It [the effective tax rate] cannot be 5% because the plan calls for a 6% tax. Yes, it can be 6% because that's what it says in the problem." This finding suggests that students had trouble deciphering critical information presented in the problem, perhaps because it differed substantially from their everyday experiences with tax rates such as sales tax.

Students' performance on algebraic reasoning items suggests that they are not well prepared to construct explanations involving their reasoning about algebraic objects. It also suggests that more emphasis needs to be given to students' analyzing or evaluating and, ultimately developing on their own, written arguments for conjectures in algebra. This is true even for students in the most advanced mathematics courses. Also, if given more time than the five minutes available for extended constructed-response questions during the NAEP assessment, students might be able to demonstrate more complete reasoning and better communicate their explanations and justification for their reasoning.

TRIGONOMETRIC FUNCTIONS

Table 9.14 contains a summary of six NAEP items dealing with trigonometry and associated performance levels for students in grade 12. For the first five items, all in multiple-choice format, percent-correct values ranged from about 20 percent to 40 percent, which suggests that twelfth-grade students had difficulty with the trigonometric concepts and skills assessed in those items. The lowest level of performance, 7 percent correct, occurred on item 6, a regular constructed-response item. A closer look at some of the released items can offer insights into why students did so poorly on trigonometry items in NAEP.

Table 9.14 Performance on Trigonometry Items

Item Description	Percent Correct Grade 12
1. Apply trigonometric concepts to a geometry problem.	42
2. Given the lengths of the legs of a right triangle, find the cosine of a particular angle.	30
3. Identify one coordinate of a point on the unit circle.	28
4. Complete a trigonometric identity.	25
5. Identify one coordinate of a point on the graph of a trigonometric function.	19
6. Apply trigonometric concepts to find the measure of an angle in a pyramid.	7

Note: Item 6 was a regular constructed-response item, and the rest were multiple-choice items.

The two released items, shown in table 9.15, required students to apply trigonometry in triangles, determining the cosine of an angle and finding the measure of a particular angle in a triangle. Although in item 1 nearly one-third of the students could find the cosine of an angle of the triangle, presumably by first finding the remaining side of the triangle, more than half of the students chose either 3/4 or 4/3. This could have resulted from confusion of cosine with cotangent; or, using the two numbers given in the problem, students simply may have guessed. On item 2 only a small percentage of twelfth-grade students correctly found an angle when one side of a triangle was given and the other could be found from the hexagon. The fact that this item was a regular constructed-response item that involved a three-dimensional figure and required an understanding of the characteristics of a regular hexagon may have made this item difficult for students.

Item 2 was classified by NAEP as calculator active, and while working on that item students had access to a scientific calculator. Of those students who indicated that they used the calculator on this item, 22 percent responded correctly. In contrast, of those students who indicated that they did not use the calculator on this item, *none* answered correctly. Here, as on other calculator items, one cannot simply attribute superior performance to calculator use. NAEP results provide no information about whether students who would have obtained correct responses were more inclined to use the calculator or whether students who chose to use the calculator were more likely to get correct responses as a result of calculator use. The most likely way in which the calculator could have helped was for computing the inverse tangent of 15/12, a value that is difficult to find without a calculator.

Although the six items pertaining to trigonometry offer a glimpse of students' knowledge of basic trigonometric definitions and identities and application of trigonometric functions to right triangles, these items were not designed to provide information about the function concept in the context of the trigonometric functions. In particular, from these items little can be concluded about students' understanding of periodicity, amplitude, domain and range, and functions and their inverses, as well as connections among symbolic, numeric, and graphic representations of trigonometric functions.

OVERALL PERFORMANCE IN ALGEBRA AND FUNCTIONS

Reports of NAEP results included looking at performance across the entire set of algebra and functions items according to the NAEP scale. Table 9.16 contains data on the average performance on this set of items for students at each grade level and for selected subgroups of students for 1990

Table 9.15 Trigonometry Items

Item	Percent Responding Grade 12

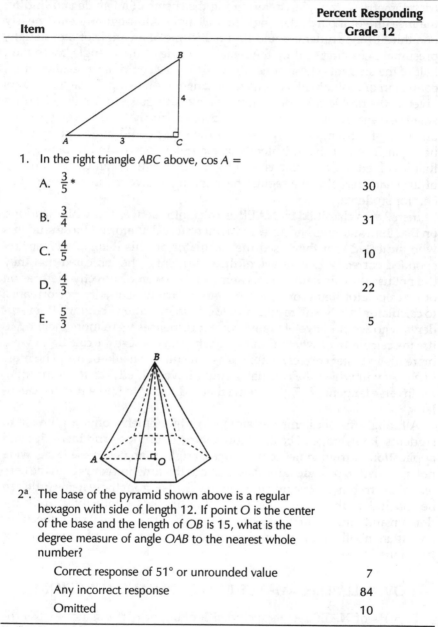

1. In the right triangle *ABC* above, cos *A* =

   A. $\frac{3}{5}$ *   30

   B. $\frac{3}{4}$ 31

   C. $\frac{4}{5}$ 10

   D. $\frac{4}{3}$ 22

   E. $\frac{5}{3}$ 5

2[a]. The base of the pyramid shown above is a regular hexagon with side of length 12. If point *O* is the center of the base and the length of *OB* is 15, what is the degree measure of angle *OAB* to the nearest whole number?

Correct response of 51° or unrounded value	7
Any incorrect response	84
Omitted	10

*Indicates correct response.

[a]Item is from an item block for which students were provided with, and permitted to use, scientific calculators.

Note: Percents may not add to 100 because of rounding or omissions.

Table 9.16 Average Performance in Algebra and Functions: Overall Performance and Performance by Gender and Race/Ethnicity, Grades 4, 8, and 12 for 1990 and 1992

	Grade 4	Grade 8	Grade 12
Overall			
1992	217 (0.9)>	267 (1.0)>	300 (1.0)>
1990	214 (0.9)	261 (1.2)	296 (1.2)
Gender			
Male			
1992	216 (1.1)	266 (1.2)>	300 (1.2)
1990	214 (1.3)	261 (1.6)	297 (1.4)
Female			
1992	217 (1.6)	269 (1.2)>	299 (1.1)>
1990	214 (1.1)	262 (1.3)	295 (1.3)
Race/Ethnicity			
White			
1992	225 (1.0)>	276 (1.2)>	305 (1.0)>
1990	221 (1.1)	269 (1.4)	302 (1.3)
Black			
1992	190 (1.6)	238 (2.1)	277 (2.1)
1990	191 (1.8)	239 (2.6)	272 (2.0)
Hispanic			
1992	198 (1.7)	245 (1.4)	283 (1.8)
1990	197 (2.2)	243 (2.9)	278 (2.8)

> The value for 1992 was significantly higher than the value for 1990 at about the 95 percent confidence level. The standard errors of the estimated proficiencies appear in parentheses.

and 1992. At all three grade levels, performance increased significantly from 1990 to 1992, with a sharper increase between grades 4 and 8 than between grades 8 and 12. Data from the NAEP teacher questionnaires for those two grades reveal that emphasis on concepts in this content area increases from the elementary grades through middle school. In particular, for 1992, although about one-third of fourth-grade students had teachers who reported giving heavy or moderate emphasis to topics in algebra and functions, by grade 8 more than 80 percent of students in that grade had teachers who emphasized this content area.

Performance increased from 1990 to 1992 for both males and females at grade 8 and females only at grade 12. When males and females were compared within a grade level in 1992, performance levels in algebra and functions were nearly identical for all three grade levels and nearly identical to the overall performance levels. For race/ethnicity subgroups, only White students at all three grades had significantly higher performance levels in 1992 than in 1990. Among the subgroups in 1992, White students outperformed both Black and Hispanic students at all three grade levels, and Hispanic students outperformed Black students at grade 4.

CONCLUSION

There is a substantial emphasis on algebra in fourth-grade NAEP items, and some younger students exhibit capabilities with informal algebra, particularly solving number sentences and describing patterns and relationships in tables. These capabilities have the potential to provide a sound basis for later formal work with variables, equations, and functions. However, generalizing and communicating to others about their reasoning is quite difficult for most students. It appears that for most students the structural aspects of algebra are not well developed, which is consistent with Kieran's (1992) conclusions. In addition, students are not well equipped for algebra if that algebra is refocused on mathematical modeling (Framework for Algebra Committee 1994).

Although quite a few NAEP items addressed traditional topics such as simplifying, solving, and graphing, some items required real-world problem solving, algebraic reasoning, and communication. It appears that the NAEP mathematics assessment has the potential to assess the type of mathematics advocated in the *Curriculum and Evaluation Standards for School Mathematics* (NCTM 1989) if items engage students in more-substantial mathematical tasks and provide students with appropriate technological tools such as graphing calculators. For example, the extended constructed-response items may afford rich insights into students' use of algebra in problem solving, their algebraic reasoning, their communication about algebra and functions, and the connections they make within algebra and between algebra and other areas of mathematics. However, it appears that the structure of the 1992 mathematics assessment did not provide sufficient time for students to demonstrate capabilities on these more complex tasks. Finally, as more students use graphing calculators, it will be important to assess students' understanding of mathematics in the presence of those tools, particularly when they solve problems related to algebra and functions.

REFERENCES

Clement, John. "Misconceptions in Graphing." In *Proceedings of the Ninth International Conference for the Psychology of Mathematics Education,* edited by Leen Streefland, pp. 369–75. Noordwijkerhout, Netherlands: State University of Utrecht, 1985.

Dossey, John A., Ina V. S. Mullis, and Chancey O. Jones. *Can Students Do Mathematical Problem Solving? Results from Constructed-Response Questions in NAEP's 1992 Mathematics Assessment.* Washington, D.C.: National Center for Education Statistics, 1993.

Framework for Algebra Committee. *Algebra for Everyone: More than a Change in Enrollment Patterns.* Reston, Va.: National Council of Teachers of Mathematics, 1994.

Kieran, Carolyn. "The Learning and Teaching of School Algebra." In *Handbook of Research on Mathematics Teaching and Learning,* edited by Douglas A. Grouws, pp. 390–419. New York: Macmillan Publishing Co., 1992.

Markovits, Zvia, Bat-Sheva Eylon, and Maxim Bruckheimer. "Functions—Linearity Unconstrained." In *Proceedings of the Seventh International Conference for the Psychology of Mathematics Education,* edited by R. Hershkowitz, pp. 271–77. Rehovot, Israel: Weizmann Institute of Science, 1983.

National Assessment of Educational Progress. *Mathematics Objectives: 1990 Assessment.* Princeton, N.J.: Educational Testing Service, National Assessment of Educational Progress, 1988.

National Council of Teachers of Mathematics. *Curriculum and Evaluation Standards for School Mathematics.* Reston, Va.: National Council of Teachers of Mathematics, 1989.

Sfard, Anna. "On the Dual Nature of Mathematical Conceptions: Reflections on Processes and Objects as Different Sides of the Same Coin." *Educational Studies in Mathematics* 22 (1991): 1–36.

Steen, Lynn A., ed. *On the Shoulders of Giants: New Approaches to Numeracy.* Washington, D.C.: National Academy Press, 1990.

Wagner, Sigrid, and Carolyn Kieran. "An Agenda for Research on the Learning and Teaching of Algebra." In *Research Issues in the Learning and Teaching of Algebra,* edited by Sigrid Wagner and Carolyn Kieran, pp. 220–37. Reston, Va.: National Council of Teachers of Mathematics, 1989.

10

Learning from NAEP: Looking Back and Looking Ahead

Edward A. Silver

THERE ARE TWO general perspectives from which to view the *message* communicated by the findings of the sixth NAEP mathematics assessment regarding the state of mathematical proficiency of the nation's students. One can view the results as transmitting a positive message by noting, for example, that students performed relatively better in 1992 than in 1990. In fact, there were statistically significant increases in average mathematics proficiency between the 1990 and 1992 NAEP assessments at all three grade levels (Mullis et al. 1993). On the other hand, one can find a negative message in the fact that the absolute level of student performance was very low in 1992, with a very small percent of the sample at each grade level demonstrating success with the most complex tasks on the NAEP assessment (see chapter 2, table 2.4). Thus, it is not surprising that when the 1992 NAEP mathematics assessment results were reported in national newspapers, one headline heralded, "U.S. Students Gain in Math for a Change," whereas another announced, "Students' Math Skills Fail Grade." Moreover, even if the contradictory messages were not present, this approach to reporting NAEP results through headlines and "sound bites" is quite limited. Such an approach is far less likely to reveal the important information about students, teachers, and schools that is embedded in the NAEP database than is the kind of careful and complete examination of findings presented in this book.

In the previous chapters various authors have discussed key findings from the 1992 NAEP mathematics assessment, which investigated the mathematical knowledge and performance of a nationally representative sample of U.S. students. They have summarized information concerning the performance of all students and of students in various demographic subgroups, and they have examined student performance on individual items and clusters of items related to specific mathematical content topics. They have also presented other information available from the NAEP mathematics assessment regarding students' attitudes and beliefs; students' enrollment in mathematics courses in the secondary school; and teachers' instructional practices, preparation, and support. These chapters have been prepared for this volume because of a shared belief among the authors that this information can be a valuable resource for teachers, teacher educators, and educational researchers interested in students' mathematics achievement.

NAEP can serve as a source of learning in several different ways: as a rich source of information about the mathematical proficiency of the nation's students, as an indicator of the school and classroom context for the observed student achievement, and as a source of assessment tasks that can be used or adapted for use in teachers' classrooms. Moreover, in this time of great interest in mathematics education reform, the findings of the 1992 NAEP mathematics assessment serve as an important marker of the state of students' achievement and the nature of school policies and classroom instruction early in the period of reform. As data become available from the 1996 NAEP assessment and from those administered in subsequent years, NAEP will serve as an important indicator of the progress of mathematics education reform in the United States.

LOOKING BACK

Some of the findings discussed in this volume dealing with the sixth NAEP mathematics assessment are quite similar to those reported for prior assessments. For example, students appeared to do better on tasks that involve basic content and contexts and on tasks that made relatively simple cognitive demands. As has been found in previous NAEP assessments, by the time they reach eighth grade, students have fairly good facility with whole-number concepts and operations, but they have far less mastery of concepts and operations associated with rational numbers (e.g., fractions, decimals, and percents), especially when the performance setting was not among the most familiar ones encountered in textbooks. Despite concerns expressed by many commentators about the possible negative effects on student performance of classroom use of calculators and other features of reform-oriented mathematics instruction, the 1992 NAEP results do not suggest any deterioration in the performance of the nation's students on test items that assess basic knowledge and skills. In fact, student performance on NAEP's long-term trend items, which assess basic and traditional mathematics concepts and skills (Mullis et al. 1994), was significantly better in 1992 than it was in 1973 for 9-year-old and 13-year-old students, but it was not significantly different from 1973 for 17-year-old students (see chapter 3, table 3.5).

Although the 1992 NAEP findings do not sound alarms regarding the need for greater attention to basic skills, the findings do suggest a need for increased vigilance in attending to higher-level goals and objectives. In the sixth NAEP mathematics assessment, as with all previous NAEP assessments, students performed better on simple, one-step problem solving than on more-complex, multistep problem solving. The need for sustained instructional attention to more-complex performance appears to be as much needed today as when it was called for by another group of mathematics educators fifteen years ago, after they had analyzed the results of the second NAEP mathematics assessment for a similar volume of interpretive reports (Carpenter et al. 1981, pp. 146–47):

> NAEP results showed that the majority of students at all age levels had difficulty with any nonroutine problem that required some analysis or thinking. ... The assessment results indicate that the primary area of concern should not be with simple one-step verbal problems, but with nonroutine problems that require more than a simple application of a single arithmetic operation. Part of the cause of students' difficulty with nonroutine problems may lie in our overemphasis on one-step problems that can be solved by simply adding, subtracting, multiplying or dividing. ... Instruction that reinforces this simplistic approach to problem solving may contribute to students' difficulty in solving unfamiliar problems. ... Students need to learn how to analyze problem situations through instruction that encourages them to think about problems and helps them to develop good problem-solving strategies.

The need for instructional activities that foster in students an ability to give sustained attention to complex mathematical problems is further fortified in the comments of the authors of chapters 5–9, who noted that students appeared to have difficulty with tasks that required the use of mathematical reasoning, and in findings of weak student performance on extended constructed-response tasks, which were included for the first time in the 1992 NAEP mathematics assessment. As Dossey and Mullis noted in chapter 2, students at all grade levels generally performed extremely poorly on NAEP's extended constructed-response tasks, which required students to solve reasonably complex mathematics problems and to explain key aspects of their solution (e.g., justify their answer or produce an elaborated solution or representation). On average, only about one in six students was able to produce a response that was judged to be satisfactory or better on the extended constructed-response tasks at grade 4; on the extended constructed-response tasks administered at grades 8 and 12, only about one in ten students was able to produce a response that was judged to be at least satisfactory (see chapter 2, table 2.4). Although the reasons for poor performance varied across tasks and grade levels (see, for example, the analysis of responses to extended examples in chapters 5, 8, and 9), one common thread running through all the analyses was the ineffectiveness of students in providing clear explanations tied to mathematical ideas. Moreover, large numbers of students at each grade level did not respond to these tasks. The large number of students who did not attempt the tasks—and the poor quality of the responses obtained from those who did attempt them—suggests that students were unfamiliar with the requirements of such tasks and that they need to have more exposure in classroom instruction to tasks that require demonstrated solutions and extended explanations or justifications.

Although the overall rate of success on the extended constructed-response tasks was considerably lower than one would desire, the performance of Black and Hispanic students and students attending schools in disadvantaged urban communities was even lower. Typically only about one in twenty students in these demographic subgroups at each grade level gave a response to these tasks that was judged to be satisfactory or better (see chapter 3, table 3.6). Unfortunately, these findings regarding the differential performance of White, Black, and Hispanic students on the NAEP extended constructed-response tasks are not an isolated phenomenon. In fact, there was a clear pattern of differential performance among White, Black, and Hispanic students across the entire 1992 NAEP mathematics assessment. And the performance differences related to socioeconomic status appear to be even greater than those associated with race/ethnicity. (See chapter 3 for a detailed discussion of these results.) Moreover, the performance differences mirror to a great extent differences in patterns of mathematics course enrollment for these subgroups of students, with Black and Hispanic students being far less likely than their White counterparts to enroll in algebra in grade 8 or to enroll in college preparatory courses in secondary school.

The findings from the 1992 NAEP mathematics assessment suggest that for Black and Hispanic students and for students attending schools in economically impoverished communities, there appears to be a critical need for sustained attention to mathematical problem solving and complex mathematical reasoning. The long-term trend data available from NAEP indicate that Black and Hispanic students have made considerable progress over the past twenty years in closing the gap with White students on items that assess basic knowledge and skills (see chapter 3, table 3.5), but the gap on more-complex tasks remains (see chapter 3, table 3.6). As more-complex tasks become more prevalent on mathematics assessments, it is likely that the performance gap could widen again.

Unfortunately, other NAEP data indicate that mathematics teachers of Black and Hispanic students at grades 4 and 8 are far more likely than mathematics teachers of White students to use multiple-choice testing at least once each week to assess the mathematics achievement of their students (see chapter 3, table 3.12). Assuming that the frequent use of this limited form of assessment reflects reasonably well other aspects of classroom instruction, then there is little reason to believe that these teachers will be able both to continue their current mathematics instructional practices and to succeed in fostering in their students the kinds of mathematical proficiency needed to perform well on NAEP extended constructed-response tasks in the future. And the generally poor performance of the nation's students on the more challenging tasks in the assessment suggests that the challenges are not isolated to Black and Hispanic students or schools in poor communities. If we are to achieve better performance on complex problem-solving tasks by *all* students, then it seems clear that fundamental changes are needed in classroom instruction and assessment.

In chapter 4 Lindquist discussed the findings of the sixth NAEP mathematics assessment regarding teachers' self-reported instructional practice. As Lindquist noted, the data support a view of mathematics classrooms in which teaching relies heavily on lectures and textbooks, in which communication on the part of students is not a salient component, and in which classroom testing (which is frequent) rarely requires more than demonstrating well-rehearsed skills. These characteristics of mathematics teaching are consistent with the poor performance of students on NAEP tasks requiring reasoning and extended responses. It is doubtful that student performance on the more complex forms of mathematics performance will improve substantially unless and until there is a corresponding change in classroom instructional practice.

Guidance for the kinds of instructional changes that may be needed can be found in the NCTM *Curriculum and Evaluation Standards for School Mathematics* (1989), the *Professional Standards for Teaching Mathematics* (1991), and the *Assessment Standards for School Mathematics* (1995). And beyond providing an indicator of teachers' classroom practice and students' mathematical performance at various time points, NAEP may also be able to contribute in another way to this effort.

If used prudently, NAEP assessment tasks may serve as a resource to teachers as they try to make a transformation in their instructional and assessment practices. For example, teachers may find it useful to use in their classroom some of the released tasks from the sixth NAEP mathematics assessment, especially the regular and extended constructed-response tasks, which require more-complex responses. These released tasks could be used as assessment tasks in their current form, in which case teachers at appropriate grade levels perform informal comparisons of their students' performance with that of NAEP's national sample. Alternatively, NAEP tasks could be adapted in appropriate ways to serve as more-complex assessment tasks or as tasks that are integrated into instruction. For example, Kenney and Silver (1997) discuss released tasks from the 1992 NAEP mathematics assessment that dealt with patterns and algebraic thinking in grade 4. In addition to presenting NAEP findings regarding student performance,

Kenney and Silver also suggest ways in which teachers might adapt some of the tasks to their instruction so their students can develop the foundation for algebraic thinking in elementary school. Likewise, other suggestions suitable for use by classroom teachers are sprinkled throughout this book, especially in chapters 5–9.

Although the results of the sixth NAEP mathematics assessment suggest that there is much that remains to be done in order to augment the mathematical proficiency of the nation's students, especially students of color and those living in poverty, the 1992 NAEP results also suggest a reason to be optimistic that success can ultimately be attained. Although the gap among race/ethnicity and socioeconomic subgroups appears persistent, the NAEP findings suggest that the gender gap has substantially closed. In particular, the 1992 NAEP results indicate that male and female student enrollment in mathematics courses in high school is now very similar, except for a slightly higher enrollment for males in calculus. And there was no significant difference in overall performance between males and females at any grade level or in any content area tested by NAEP. In the past twenty years a variety of efforts have been undertaken to heighten awareness of, and design remedies for, gender inequity in mathematics education, and the 1992 NAEP results suggest that these efforts have paid off in measurable ways. Thus, there is reason to be optimistic that a sustained focus on mathematics education inequities associated with race, ethnicity, and socioeconomic status could have similar efficacy.

LOOKING AHEAD

Not only with respect to equity initiatives but more generally with respect to efforts to reform the content and character of mathematics education in the nation's schools, the results of the sixth NAEP mathematics assessment can be seen as providing baseline information about the state of students' mathematics proficiency and the associated classroom and school context in which that proficiency develops. As was pointed out in chapter 1, the framework that guided the development of the fifth and sixth NAEP mathematics assessments, which were administered in 1990 and 1992, was created at approximately the same time as the NCTM *Curriculum and Evaluation Standards* (1989). Therefore, the results of these assessments afford a glimpse of the conditions at about the time when the reform journey began. The results of the seventh NAEP mathematics assessment, which was administered in 1996, should provide more information about the progress of the early part of the reform journey.

A new NAEP mathematics framework (National Assessment Governing Board 1996) was used to guide the development of the seventh NAEP mathematics assessment, and it contained important changes from the framework used to create the 1990 and 1992 NAEP mathematics tests. Some of the most important design features for the 1996 NAEP in mathematics include the following:

- Abandoning a rigid content-by-process matrix framework employed in the 1990 and 1992 NAEP assessments, with specified percentages of items in each row and column of the matrix, in favor of building the assessment primarily around five major content strands (similar to those used in the 1990 and 1992 assessments), with attention to assuring a reasonable balance among mathematical processes

- Allowing the classification of a single item in more than one content strand and/or ability level if the item warrants such multiple classification

- Modifying the role of the mathematics ability categories (Conceptual Understanding, Procedural Knowledge, and Problem Solving) so that they no longer define specific percentages of items in each of the five content strands but rather are used along with the process goals of reasoning, connections, and communication to define item descriptors and to achieve a balance across the task sets for each grade level

- Continuing the effort begun with the 1990 NAEP mathematics assessment to create an assessment that balances attention to all content strands, rather than being dominated by numbers and operations

- Increasing the number of constructed-response items as much as possible within the bounds of the statistical design used to carry out the assessment

- Creating "families" of tasks, that is, sets of related assessment items that probe students' mathematical understandings and proficiencies vertically (within a content area) or horizontally (across content areas)

This list of design features for the 1996 NAEP mathematics assessment suggests that the next NAEP assessment will reflect the vision for school mathematics presented in the NCTM *Curriculum and Evaluation Standards for School Mathematics* more closely than did the 1992 assessment.

CODA

The information presented in this book, and in the one that will undoubtedly be written concerning the results of the seventh NAEP mathematics assessment, provides a valuable database for educational decision makers, whether in the mathematics classroom or in a state department of education. Nevertheless, the information alone does not make the important decisions that affect students. The interpretations provided by the authors of chapters in this volume can certainly help those who wish to use the NAEP data to make decisions, but even these interpretations must be considered in the light of issues and questions that reside outside of the assessment itself. The authors of the interpretive volume dealing with the first NAEP mathematics assessment (Carpenter et al. 1978, p. 138) concluded with these words, which are still important to ponder today:

> We believe that the monograph provides a data base from which to ask questions about mathematics achievement. Such activity forces recognition that some fundamental issues precede interpretation of data. What is satisfactory performance in computational skills? What uses of mathematics can be expected of all citizens? These and many other questions must be faced. National Assessment and its advisors face them in planning, in exercise development, and in reporting. But the reader also must deal with such questions as he interprets the information presented and tries to answer, "What do these data say for me, for my school, or my students?"

I would like to update this notion to the 1990s by suggesting that some key questions of this type for consideration by readers of this volume might include the following: What degree of proficiency should we expect of students on extended constructed-response tasks? How proficient should students be in deciding when and how to use calculators to solve mathematics problems? What level of student performance should be expected of all students in each of the content areas and overall? Ultimately these are questions of value, and their answers are deeply entwined with the interpretation one gives to the findings of any NAEP mathematics assessment.

In this volume the authors have shared their interpretations of the findings of the sixth NAEP mathematics assessment, taking care to provide as much information as possible directly from the assessment to help readers understand the basis for the interpretations given. The important result of this work is not that all readers agree with the interpretations provided but rather that they examine carefully the information provided, interpret the findings in their own way, and draw their own conclusions. In this way, readers can transform their concern for improving the mathematical achievement of our nation's students into policies and practices that will ensure a high-quality mathematics education for all students.

REFERENCES

Carpenter, Thomas P., Terrence G. Coburn, Robert E. Reys, and James W. Wilson. *Results from the First Mathematics Assessment of the National Assessment of Educational Progress.* Reston, Va.: National Council of Teachers of Mathematics, 1978.

Carpenter, Thomas P., Mary Kay Corbitt, Henry S. Kepner, Jr., Mary Montgomery Lindquist, and Robert E. Reys. *Results from the Second Mathematics Assessment of the National Assessment of Educational Progress.* Reston, Va.: National Council of Teachers of Mathematics, 1981.

Kenney, Patricia Ann, and Edward A. Silver. "Probing the Foundations of Algebra: Grade-4 Pattern Items in NAEP." *Teaching Children Mathematics* 3 (February 1997): 268–74.

Mullis, Ina V. S., John A. Dossey, Jay R. Campbell, Claudia A. Gentile, Christine O'Sullivan, and Andrew S. Latham. *NAEP 1992 Trends in Academic Progress.* 23-TR-01. Washington, D.C.: National Center for Education Statistics, 1994.

Mullis, Ina V. S., John A. Dossey, Eugene H. Owen, and Gary W. Phillips. *NAEP 1992 Mathematics Report Card for the Nation and the States: Data from National and Trial State Assessments.* 23-ST-02. Washington, D.C.: National Center for Education Statistics, April 1993.

National Assessment Governing Board. *Mathematics Framework for the 1996 National Assessment of Educational Progress.* Washington, D.C.: National Assessment Governing Board, U.S. Department of Education, 1996.

National Council of Teachers of Mathematics. *Assessment Standards for School Mathematics.* Reston, Va.: National Council of Teachers of Mathematics, 1995.

_____. *Curriculum and Evaluation Standards for School Mathematics.* Reston, Va.: National Council of Teachers of Mathematics, 1989.

_____. *Professional Standards for Teaching Mathematics.* Reston, Va.: National Council of Teachers of Mathematics, 1991.

BIBLIOGRAPHY

SIXTH NAEP MATHEMATICS ASSESSMENT, 1992

NCES/NAEP Publications

Dossey, John A., Ina V. S. Mullis, Steven Gorman, and Andrew S. Latham. *How School Mathematics Functions: Perspectives from the NAEP 1990 and 1992 Assessments.* 23-FR-02. Washington, D.C.: National Center for Education Statistics, October 1994.

Dossey, John A., Ina V. S. Mullis, and Chancey O. Jones. *Can Students Do Mathematical Problem Solving? Results from Constructed-Response Questions in NAEP's 1992 Mathematics Assessment.* 23-FR-01. Washington, D.C.: National Center for Education Statistics, August 1993.

Johnson, Eugene, and James E. Carlson. *The NAEP 1992 Technical Report.* 23-TR-20. Washington, D.C.: National Center for Education Statistics, July 1994.

Johnson, Eugene, John Mazzeo, and Deborah L. Kline. *Technical Report of the NAEP 1992 Trial State Assessment Program in Mathematics.* Princeton, N.J.: Educational Testing Service, National Assessment of Educational Progress, 1993.

Lindquist, Mary Montgomery, John A. Dossey, and Ina V. S. Mullis. *Reaching Standards: A Progress Report on Mathematics.* Princeton, N.J.: Policy Information Center, Educational Testing Service, 1995.

Mullis, Ina V. S. *The NAEP Guide: A Description of the Content and Methods of the 1990 and 1992 Assessments.* 21-TR-01. Washington, D.C.: National Center for Education Statistics, November 1991.

Mullis, Ina V. S., ed. *America's Mathematics Problem: Raising Student Achievement.* 23-FR-03. Washington, D.C.: National Center for Education Statistics, October 1994.

Mullis, Ina V. S., John A. Dossey, Jay R. Campbell, Claudia A. Gentile, Christine O'Sullivan, and Andrew S. Latham. *NAEP 1992 Trends in Academic Progress.* 23-TR-01. Washington, D.C.: National Center for Education Statistics, July 1994.

Mullis, Ina V. S., John A. Dossey, Eugene H. Owen, and Gary W. Phillips. *NAEP 1992 Mathematics Report Card for the Nation and the States: Data from the National and Trial State Assessments.* 23-ST-02. Washington, D.C.: National Center for Education Statistics, April 1993.

Mullis, Ina V. S., Frank Jenkins, and Eugene G. Johnson. *Effective Schools in Mathematics: Perspectives from the NAEP 1992 Assessment.* 23-RR-01. Washington, D.C.: National Center for Education Statistics, October 1994.

National Center for Education Statistics. *Data Compendium for the NAEP 1992 Mathematics Assessment of the Nation and the States.* 23-ST-04. Washington, D.C.: National Center for Education Statistics, May 1993.

NCTM Publications

Kenney, Patricia Ann, and Edward A. Silver, eds. *Results from the Sixth Mathematics Assessment of the National Assessment of Educational Progress.* Reston, Va.: National Council of Teachers of Mathematics, 1997.

Other Publications

Kenney, Patricia Ann. "A Framework for the Qualitative Analysis of Student Responses to the Extended Constructed-Response Questions from the 1992 NAEP in Mathematics." In *Proceedings of the Seventeenth Annual Meeting of the North American Chapter of the International Group for the Psychology of Mathematics Education,* vol. 1, edited by Douglas. T. Owens, M. K. Reed, and G. M. Millsaps, pp. 175–80. Columbus, Ohio: ERIC Clearinghouse for Science, Mathematics, and Environmental Education, 1995.

Silver, Edward A., and Patricia Ann Kenney. "The Content and Curricular Validity of the 1992 NAEP TSA in Mathematics." In *The Trial State Assessment: Prospects and Realities: Background Studies,* pp. 231–84. Stanford, Calif.: National Academy of Education, 1994.

———. "Expert Panel Review of the 1992 NAEP Mathematics Achievement Levels." In *Setting Performance Standards for Student Achievement: Background Studies,* pp. 215–81. Stanford, Calif.: National Academy of Education, 1993.

———. *Understanding Students' Mathematical Problem Solving: A Commentary on the 1992 NAEP Findings.* Pittsburgh, Pa.: Learning Research and Development Center, University of Pittsburgh, 1994.

FIFTH NAEP MATHEMATICS ASSESSMENT, 1990

NCES/NAEP Publications

Johnson, Eugene G., and Nancy L. Allen. *The NAEP 1990 Technical Report.* 21-TR-20. Princeton, N.J.: Educational Testing Service, National Assessment of Educational Progress, 1992.

Mullis, Ina V. S., John A. Dossey, Eugene H. Owen, and Gary W. Phillips. *The STATE of Mathematics Achievement: NAEP's 1990 Assessment of the Nation and the Trial Assessment of the States.* 21-ST-04. Washington, D.C.: National Center for Education Statistics, June 1991.

Mullis, Ina V. S., Eugene H. Owen, and Gary W. Phillips. *America's Challenge: Accelerating Academic Achievement: A Summary of Findings from 20 Years of NAEP*. 19-OV-01. Washington, D.C.: National Center for Education Statistics, September 1990.

National Assessment of Educational Progress. *Mathematics Objectives: 1990 Assessment*. 21-M-10. Princeton, N.J.: Educational Testing Service, National Assessment of Educational Progress, 1988.

NCTM Publications

Silver, Edward A., and Patricia Ann Kenney. "An Examination of Relationships between the 1990 NAEP Mathematics Items for Grade 8 and Selected Themes from the NCTM Standards." *Journal for Research in Mathematics Education* 24 (March 1993): 159–67.

Other Publications

Silver, Edward A., Patricia Ann Kenney, and Leslie Salmon-Cox. "The Content and Curricular Validity of the 1990 NAEP Mathematics Items: A Retrospective Analysis." In *Assessing Student Achievement in the States: Background Studies*, pp. 157–218. Stanford, Calif.: National Academy of Education, 1992.

FOURTH NAEP MATHEMATICS ASSESSMENT, 1985–86

NAEP Publications

Beaton, Albert E. *Implementing the New Design: The NAEP 1983–84 Technical Report*. 15-TR-20. Princeton, N.J.: Educational Testing Service, March 1987.

Dossey, John A., Ina V. S. Mullis, Mary M. Lindquist, and Donald L. Chambers. *The Mathematics Report Card: Are We Measuring Up?* 17-M-01. Princeton, N.J.: Educational Testing Service, 1988.

National Assessment of Educational Progress. *Math Objectives: 1985–86 Assessment*. 17-M-10. Princeton, N.J.: Educational Testing Service, National Assessment of Educational Progress, 1986.

NCTM Publications

Brown, Catherine A., Thomas P. Carpenter, Vicky L. Kouba, Mary M. Lindquist, Edward A. Silver, and Jane O. Swafford. "Secondary School Results for the Fourth NAEP Mathematics Assessment: Algebra, Geometry, Mathematical Methods, and Attitudes." *Mathematics Teacher* 81 (May 1988): 337–47, 397.

_____. "Secondary School Results for the Fourth NAEP Mathematics Assessment: Discrete Mathematics, Data Organization and Interpretation, Measurement, Number and Operations." *Mathematics Teacher* 81 (April 1988): 241–48.

Carpenter, Thomas P., Mary M. Lindquist, Catherine A. Brown, Vicky L. Kouba, Edward A. Silver, and Jane O. Swafford. "Results of the Fourth NAEP Assessment of Mathematics : Trends and Conclusions." *Arithmetic Teacher* 36 (December 1988): 38–41.

Kouba, Vicky L., Catherine A. Brown, Thomas P. Carpenter, Mary M. Lindquist, Edward A. Silver, and Jane O. Swafford. "Results of the Fourth NAEP Assessment of Mathematics: Measurement, Geometry, Data Interpretation, Attitudes, and Other Topics." *Arithmetic Teacher* 35 (May 1988): 10–16.

_____. "Results of the Fourth NAEP Assessment of Mathematics: Number, Operations, and Word Problems." *Arithmetic Teacher* 35 (April 1988): 14–19.

Lindquist, Mary Montgomery, ed. *Results from the Fourth Mathematics Assessment of the National Assessment of Educational Progress.* Reston, Va.: National Council of Teachers of Mathematics, 1989.

Silver, Edward A., Mary M. Lindquist, Thomas P. Carpenter, Catherine A. Brown, Vicky L. Kouba, and Jane O. Swafford. "The Fourth NAEP Mathematics Assessment: Performance Trends and Results and Trends for Instructional Indicators." *Mathematics Teacher* 81 (December 1988): 720–27.

Other Publications

Lindquist, Mary M., Thomas P. Carpenter, Catherine A. Brown, Vicky L. Kouba, Edward A. Silver, and Jane O. Swafford. "NAEP: Results of the Fourth Mathematics Assessment." *Education Week*, 15 June 1988, 28–29.

THIRD NAEP MATHEMATICS ASSESSMENT, 1981–82

NAEP Publications

National Assessment of Educational Progress. *Mathematics Objectives, 1981–82 Assessment.* Denver: Education Commission of the States, 1981.

_____. *The Third National Mathematics Assessment: Results, Trends, and Issues.* 13-MA-01. Denver: Education Commission of the States, 1981.

NCTM Publications

Carpenter, Thomas P., Mary M. Lindquist, Westina Matthews, and Edward A. Silver. "Results of the Third NAEP Mathematics Assessment: Secondary School." *Mathematics Teacher* 76 (December 1983): 652–59.

Lindquist, Mary Montgomery, Thomas P. Carpenter, Edward A. Silver, and Westina Matthews. "The Third National Mathematics Assessment: Results and Implications for Elementary and Middle Schools." *Arithmetic Teacher* 31 (December 1983): 14–19.

Matthews, Westina, Thomas P. Carpenter, Mary Montgomery Lindquist, and Edward A. Silver. "The Third National Assessment: Minorities and Mathematics." *Journal for Research in Mathematics Education* 15 (March 1984): 165–71.

Other Publications

Carpenter, Thomas P., Westina Matthews, Mary Montgomery Lindquist, and Edward A. Silver. "Achievement in Mathematics: Results from the National Assessment." *Elementary School Journal* 84 (May 1984): 485–97.

SECOND NAEP MATHEMATICS ASSESSMENT, 1977–78

NAEP Publications

National Assessment of Educational Progress. *Changes in Mathematical Achievement: 1973–78.* 09-MA-01. Denver: Education Commission of the States, 1979.

_____. *Mathematical Applications.* 09-MA-03. Denver: Education Commission of the States, 1979.

_____. *Mathematical Knowledge and Skills.* 09-MA-02. Denver: Education Commission of the States, 1979.

_____. *Mathematical Objectives: Second Assessment.* Denver: Education Commission of the States, 1978.

_____. *Mathematical Understanding.* 09-MA-04. Denver: Education Commission of the States, 1979.

_____. *The Second Assessment of Mathematics: 1977–78: Released Exercise Set.* Denver: Education Commission of the States, 1979.

NCTM Publications

Anick, Constance Martin, Thomas P. Carpenter, and Carol Smith. "Minorities and Mathematics: Results from the National Assessment." *Mathematics Teacher* 74 (October 1981): 560–66.

Bestgen, Barbara J. "Making and Interpreting Graphs and Tables: Results and Implications from National Assessment." *Arithmetic Teacher* 28 (December 1980): 26–29.

Carpenter, Thomas P., Mary Kay Corbitt, Henry S. Kepner, Jr., Mary Montgomery Lindquist, and Robert E. Reys. "Calculators in Testing Situations: Results and Implications from National Assessment." *Arithmetic Teacher* 28 (January 1981): 34–37.

_____. "The Current Status of Computer Literacy: NAEP Results for Secondary Students." *Mathematics Teacher* 73 (December 1980): 669–73.

_____. "Decimals: Results and Implications from National Assessment." *Arithmetic Teacher* 28 (April 1981): 34–37.

_____. "NAEP Note: Problem Solving." *Mathematics Teacher* 73 (September 1980): 427–33.

_____. "Results and Implications of the Second NAEP Mathematics Assessments: Elementary School." *Arithmetic Teacher* 27 (April 1980): 10–12, 44–47.

_____. *Results from the Second Mathematics Assessment of the National Assessment of Educational Progress.* Reston, Va.: National Council of Teachers of Mathematics, 1981.

_____. "Results of the Second NAEP Mathematics Assessment: Secondary School." *Mathematics Teacher* 73 (May 1980): 329–38.

_____. "Solving Verbal Problems: Results and Implications from National Assessment." *Arithmetic Teacher* 28 (September 1980): 8–12.

_____. "Students' Affective Responses to Mathematics: Results and Implications from National Assessment." *Arithmetic Teacher* 28 (October 1980): 34–37, 52–53.

_____. "Students' Affective Responses to Mathematics: Secondary School Results from National Assessment." *Mathematics Teacher* 73 (October 1980): 531–39.

_____. "What Are the Chances of Your Students Knowing Probability?" *Mathematics Teacher* 74 (May 1981): 342–44.

Fennema, Elizabeth, and Thomas P. Carpenter. "Sex-Related Differences in Mathematics: Results from National Assessment." *Mathematics Teacher* 74 (October 1981): 554–59.

Hiebert, James J. "Units of Measure: Results and Implications from National Assessment." *Arithmetic Teacher* 28 (February 1981): 38–43.

Hirstein, James J. "The Second National Assessment in Mathematics: Area and Volume." *Mathematics Teacher* 74 (December 1981): 704–8.

Kerr, Donald R., Jr. "A Geometry Lesson from National Assessment." *Mathematics Teacher* 74 (January 1981): 27–32.

McKillip, William D. "Computational Skill in Division: Results and Implications from National Assessment." *Arithmetic Teacher* 28 (March 1981): 34–37.

Post, Thomas R. "Fractions: Results and Implications from National Assessment." *Arithmetic Teacher* 28 (May 1981): 26–31.

Rathmell, Edward C. "Concepts of the Fundamental Operations: Results and Implications from National Assessment." *Arithmetic Teacher* 28 (November 1980): 34–37.

Other Publications

Carpenter, Thomas P., Mary Kay Corbitt, Henry S. Kepner, Jr., Mary Montgomery Lindquist, and Robert E. Reys. "An Interpretation of the Results of the Second NAEP Mathematics Assessment." In *Education in the 80s: Mathematics*, edited by Shirley Hill. Washington, D.C.: National Education Association, 1982.

_____. "National Assessment: Implications for the Curriculum of the 1980s." In *Research in Mathematics Education: Implications for the 80s*, edited by Elizabeth Fennema. Washington, D.C.: Association for Supervision and Curriculum Development, 1981.

_____. "A Perspective of Students' Mastery of Basic Skills." In *Selected Issues in Mathematics Education*, edited by Mary Montgomery Lindquist. Chicago: National Society for the Study of Education, 1980.

_____. "Problem Solving in Mathematics: National Assessment Results." *Educational Leadership* 37 (April 1980): 562–63.

FIRST NAEP MATHEMATICS ASSESSMENT, 1972–73
NAEP Publications

National Assessment of Educational Progress. *Consumer Math: Selected Results from the First National Assessment of Mathematics.* 04-MA-02. Denver: Education Commission of the States, June 1975.

_____. *The First National Assessment of Mathematics: An Overview.* 04-MA-00. Denver: Education Commission of the States, October 1975.

_____. *Math Fundamentals: Selected Results from the First National Assessment of Mathematics.* 04-MA-01. Denver: Education Commission of the States, January 1975.

_____. *Mathematics Objectives.* Ann Arbor, Mich.: National Assessment of Educational Progress, 1970.

_____. *Mathematics Technical Report: Exercise Volume.* 04-MA-20. Denver: Education Commission of the States, February 1977.

_____. *Mathematics Technical Report: Summary Volume.* 04-MA-21. Denver: Education Commission of the States, September 1976.

NCTM Publications

Carpenter, Thomas P., Terrence G. Coburn, Robert E. Reys, and James W. Wilson. "Notes from National Assessment: Addition and Multiplication with Fractions." *Arithmetic Teacher* 23 (February 1976): 137–42.

_____. "Notes from National Assessment: Basic Concepts of Area and Volume." *Arithmetic Teacher* 22 (October 1975): 501–7.

_____. "Notes from National Assessment: Estimation." *Arithmetic Teacher* 23 (April 1976): 296–302.

_____. "Notes from National Assessment: Perimeter and Area." *Arithmetic Teacher* 22 (November 1975): 586–90.

_____. "Notes from National Assessment: Processes Used on Computational Exercises." *Arithmetic Teacher* 23 (March 1976): 217–22.

_____. "Notes from National Assessment: Recognizing and Naming Solids." *Arithmetic Teacher* 23 (January 1976): 62–66.

_____. "Notes from National Assessment: Word Problems." *Arithmetic Teacher* 23 (May 1976): 389–93.

_____. "Research Implications and Questions from the Year 04 NAEP Mathematics Assessment." *Journal for Research in Mathematics Education* 7 (November 1976): 327–36.

_____. "Results and Implications of the NAEP Mathematics Assessment: Elementary School." *Arithmetic Teacher* 22 (October 1975): 438–50.

_____. "Results and Implications of the NAEP Mathematics Assessment: Secondary School." *Mathematics Teacher* 68 (October 1975): 453–70.

_____. *Results from the First Mathematics Assessment of the National Assessment of Educational Progress.* Reston, Va.: National Council of Teachers of Mathematics, 1978.

_____. "Subtraction: What Do Students Know?" *Arithmetic Teacher* 22 (December 1975): 653–57.

Martin, Wayne H., and James W. Wilson. "The Status of National Assessment in Mathematics." *Arithmetic Teacher* 21 (January 1974): 49–53.

Other Publications

Reys, Robert E. "Consumer Math: Just How Knowledgeable Are U.S. Young Adults?" *Phi Delta Kappan* (November 1976): 258–60.